3/13/89

PRACTICAL HPLC METHOD DEVELOPMENT

PRACTICAL HPLC
METHOD DEVELOPMENT

Lloyd R. Snyder
LC Resources, Inc.
Orinda, California

Joseph L. Glajch and Joseph J. Kirkland
E. I. du Pont de Nemours and Company
Wilmington, Delaware

WILEY

A WILEY–INTERSCIENCE PUBLICATION

JOHN WILEY & SONS
New York · Chichester · Brisbane · Toronto · Singapore

Library of Congress Cataloging-in-Publication Data

Snyder, Lloyd R.
  Practical HPLC method development.

  "A Wiley-Interscience publication."
  1. High performance liquid chromatography—
Methodology. I. Glajch, Joseph L.   II. Kirkland, J. J.
(Joseph Jack), 1925-   . III. Title.

QP519.9.H53S69   1988   543′.0894   88-238
ISBN 0-471-62782-8

Printed in the United States of America

10 9 8 7 6 5 4 3 2 1

PREFACE

High-performance liquid chromatography (HPLC) has been in use since the late 1960s. Since that time, much has been learned about the basis of HPLC separations and how they can be improved by varying experimental conditions. This increase in basic understanding has been chronicled in numerous books, review articles, and countless other publications. However, the practical application of this knowledge is, at best, uneven. Nevertheless, a few systematic approaches to method development are now available as computer-aided packages. In some cases, all that is necessary is to load the HPLC system with a sample; the computer then takes over until a satisfactory separation has been achieved. However, the performance of present "automatic method development" schemes is somewhat erratic (1).

In most laboratories, HPLC method development proceeds empirically. Usually, reversed-phase chromatography is tried first, and the percent organic in the mobile phase is varied in an attempt to obtain adequate resolution, a reasonable run time, and easily detected (narrow) bands. Often, this approach is not successful, and then chromatographers follow their own artistic bent. Many workers try different columns; others vary temperature or mobile-phase pH. Some chromatographers try some form of solvent optimization, where the organic solvent is replaced by a different solvent. Or a new HPLC method may be attempted, such as ion-pair chromatography, normal-phase chromatography, or ion-exchange chromatography.

Getting a separation is not the only challenge encountered in method development. Initial runs often exhibit broad and/or tailing peaks. While these should be corrected before proceeding further, many chromatographers adopt a tolerant attitude to this problem. (This is a mistake.) In other cases, bands are unresolved near the beginning of the chromatogram (at t_0), while later bands exhibit excessive retention times and poor detectability. Here, gradient elution should be used, but the practicing chromatographer often prefers to defer this approach for as long as possible. Occasionally, the main challenge is detecting the separated sample bands, because of poor detector sensitivity or very low sample concentration. In this case, the chromatographer cannot avoid the problem, since the first requirement of method development is a visible chromatogram.

Fortunately, the power of HPLC is such that a reasonable separation usually can be obtained, regardless of the approach followed. Unfortunately, this may also require a good deal of time, for some practitioners as much as several months for "difficult" samples. Often, an inefficient method-development procedure leads to compromising the final method, where the quality of separation is sacrificed in the interest of reducing the time spent on method development.

It is our premise that there are better ways to develop an HPLC method. An efficient approach will include some or all of the following recommendations:

1. Begin by evaluating information on the sample and the goals of separation. (The importance of this step should not be minimized.)
2. Select conditions for the initial separation that are likely to yield a promising chromatogram in one or two runs.
3. Anticipate possible separation problems and make use of proven remedies as problems arise.
4. Prioritize the different options for controlling (and improving) separation; the most promising choices should be tried first.
5. Avoid random trial-and-error experiments. Each run should be designed to provide useful information, regardless of the quality of the resulting separation.
6. Recognize that an HPLC method should be tailored to the goals of the separation; more-than-adequate separation of bands of interest is not required, but the final method must meet the needs of the analyst. Time and effort spent on method development should be minimized, without compromising the goals of separation.
7. If possible, use a computer to organize data from initial exploratory runs so as to minimize the total number of experiments required during method development. Evaluate and select commercial software that can facilitate method development.

This book shows how to implement the above guidelines during HPLC method development. It also provides a thorough discussion of the basis for these recommendations, so that the practicing chromatographer can adapt this approach to individual problems—or override our advice where that is appropriate. We assume that the reader is already familiar with the basics of HPLC such as set forth in *An Introduction to Modern Liquid Chromatography* (2); frequent reference will be made to this text. The Glossary of Symbols and Terms at the beginning of this book should also be consulted, since these symbols are used throughout the book without introduction.

A second objective of this book is to introduce the reader to some practical tools based on the use of the computer in HPLC method development. The computer is not essential for practical method development; however, it does promote better methods and allows the chromatographer to spend less time on method development. Computerized method development has been reviewed by Berridge (3) and Schoenmakers (4), but with emphasis on state-of-the-art software that either is not commercial or forms part of an expensive HPLC system. Several computer programs that are more flexible and less costly have since become available. These programs are built on the basic theory of HPLC and are increasingly being used to more efficiently define the optimum conditions for the separation of interest.

This book does not specifically treat samples of biological origin. While the separation of such samples is governed by the same general principles that apply to all compounds, there are some practical differences that are important. Insights regarding optimum approaches for handling biological samples are developing quite rapidly at present; a unified scheme for separating such materials is not yet feasible. Therefore, we have elected to wait until a later edition of this book to discuss the special case of HPLC separations for the life sciences.

It will be apparent that our discussion of method development is directed mainly towards the reversed-phase separations of small molecules. This focus seems appropriate, since at present about three-fourths of all separations are performed by this method. However, we also address the development of normal-phase and ion-pair separations, even though systematic information on these methods is more limited. Separations by size-exclusion chromatography (SEC) are outside the scope of this book, since an entirely different set of concepts for optimizing separations is involved. Also, SEC is mainly used for separating and characterizing macromolecules, a subject that we have chosen not to address. This specialized area is adequately covered elsewhere (5).

Finally, we wish to thank Paul E. Antle, John W. Dolan, Roy Eksteen, Barbara Ewing, Sal Fusari, Matthew S. Klee, and Mark Watson for their helpful comments, and the DuPont Company for its support.

REFERENCES

1. S. A. Borman, *Anal. Chem.*, *58* (1986) 1192A.
2. L. R. Snyder and J. J. Kirkland, *Introduction to Modern Liquid Chromatography*, 2nd ed., Wiley-Interscience, New York, 1979.

3. J. C. Berridge, *Techniques for the Automated Optimization of HPLC Separations,* Wiley-Interscience, New York, 1985.

4. P. J. Schoenmakers, *Optimization of Chromatographic Selectivity,* Elsevier, Amsterdam, 1986.

5. W. W. Yau, J. J. Kirkland, and D. D. Bly, *Modern Size-Exclusion Liquid Chromatography,* Wiley, New York, 1979.

CONTENTS

GLOSSARY OF
SYMBOLS AND
TERMS

ACN	Acetonitrile
AUFS	Absorbance units, full scale
A_s	Peak asymmetry factor (see Fig. 3.7)
B	Strong solvent used in a binary-solvent mobile phase; e.g., methanol in methanol/water mixtures used in reversed-phase HPLC
BSA	Bovine serum albumin (a protein)
CAF	Caffeine (a neutral solute)
CRF	Chromatographic response function; a quantitative measure of the overall resolution of a chromatogram
d_c	Column internal diameter (cm)
DMOA	Dimethyloctylamine
DNB	2,4-Dinitrobenzoyl
d_p	Column-packing particle size (μm)
DRYLAB®	Computer-simulation software from LC Resources Inc. DRYLAB I is used for isocratic predictions, DRYLAB G for gradient predictions (see Sect. 8.4)
F	Mobile-phase flow rate (mL/min)
FC-113	1,1,2-trifluoro-1,2,2-trichloroethane
GPC	Gel-permeation chromatography
HA	an acidic solute that ionizes to give A^-
Hex	Hexane
h_r	Height of valley between two adjacent bands (see Fig. 2.8)
HVA	Homovanillic acid

h'	Peak height (arbitrary units)
h_1, h_2	Peak heights for adjacent bands 1 and 2 (see Fig. 2.8)
IEC	Ion-exchange chromatography
IP	Ion-pair
IPC	Ion-pair chromatography
J	Column-strength parameter (see Table 5.2)
k'	Capacity factor for a given band, equal to $t_0/(t_R - t_0)$
$\bar{k}$	Average or effective value of k' for a solute during gradient elution (see Fig. 6.4b and Eqn. 6.2)
k_w	Extrapolated value of k' for water as mobile phase (reversed-phase HPLC; see Eqn. 8.1)
k_1, k_2	Capacity factors k' for adjacent bands 1 and 2
L	Column length (cm)
L_c	Length of detector-flowcell lightpath (cm)
M	Molecular weight of solute
MC	Methylene chloride
MDST	Mixture-design statistical technique; software for mobile-phase optimization (see Sect. 8.4)
MeOH	Methanol
MTBE	Methyl-t-butyl ether
MW	Molecular weight of solute
N	Column plate number (see Eqn. 2.3a)
NAPA	N-Acetylprocainamide (basic solute)
N_0	Detector baseline noise (see Eqn. 4.1)
ODS	Octadecylsilyl
P	Pressure drop across column (usually in bar, but also in psi [see Eqn. 3.1]); also, column-polarity parameter (see Table 5.2)
PA	Procainamide (basic solute)
PAH	Polyaromatic hydrocarbon
PESOS®	Computer software for mobile-phase optimization (Perkin-Elmer; see Sect. 8.4)
pK_a	Negative logarithm of solute acidity constant; solute half-ionized when $pH = pK_a$
R_k	Retention range, equal to (k' for last band)/(k' for first band)
RRM	Relative resolution map (usually for $N = 10,000$; e.g., Fig. 2.10)

R_s	Resolution of two adjacent bands (see Eqns. 2.1, 2.2, 2.3)
S	Parameter that measures rate of change of solute retention as %-B in mobile phase is changed (see Eqn. 8.1)
SAL	Salicylic acid
SEC	Size-exclusion chromatography
S/N	Signal-to-noise ratio
t	Time (min) during separation (sample injected at $t = 0$)
t_D	Dwell time of gradient system (min) (see Fig. 6.10)
TBA	Tetrabutylammonium ion
TEA	Triethylamine
THF	Tetrahydrofuran
t_k	Time (min) during gradient elution when mobile phase of correct solvent strength for isocratic elution leaves the gradient mixer (see discussion of Eqn. 2.4)
TLC	Thin-layer chromatography
TMA	Tetramethylammonium (salt)
TMS	Trimethylsilyl
t_0	Column dead-time (min)
t_R	Solute retention time (min)
t_G	Gradient time (min); elapsed time between start and end of gradient
t_1, t_2	Retention times (min) of adjacent bands 1 and 2 (see Eqn. 2.1)
t_i	Retention time (min) of first peak in the chromatogram
t_f	Retention time (min) of the final (last) peak in the chromatogram
Δt_g	$t_f - t_i$
t_x	$(t_f + t_i)/2$ (see Fig. 9.6)
UV	Ultraviolet
V_m	Column dead-volume (mL), equal to $t_0 F$
VMA	Vanillylmandelic acid
w_m	Weight of compound injected (see Eqn. 4.1)
w_1, w_2	Bandwidths (min) at half-height ($W_{1/2}$) for adjacent bands 1 and 2 (see Eqn. 2.2)
W_1, W_2	Baseline bandwidths (min) of adjacent bands 1 and 2 (see Eqn. 2.1)
$W_{1/2}$	Bandwidth at half-height

x_d, x_e, x_n — Solvent-selectivity parameters that measure solvent acidity, basicity, and dipolarity, respectively (see Fig. 2.16)

α — Separation factor, equal to k_2/k_1

$\Delta\Phi$ — Change in mobile-phase composition (Φ) during gradient

ϵ° — Solvent-strength parameter (see Table 2.4)

ϵ — Molar absorptivity of a compound (see Sect. 4.1)

η — Mobile-phase viscosity (cPoise)

Φ — Volume-fraction of strong-solvent in mobile phase

%-B — Percent by volume (%v) of strong solvent in a binary-solvent mobile phase

PRACTICAL HPLC METHOD DEVELOPMENT

1

GETTING STARTED

Every day many chromatographers face the need to develop an HPLC separation. While individual approaches may exhibit considerable diversity, method development often follows the series of steps summarized in Fig. 1.1. In this chapter we will review the importance of each of these steps, in preparation for a more detailed examination in following chapters.

1.1 WHAT IS KNOWN BEFORE STARTING

Before beginning method development, we need to review what is known about the sample. The goals of the separation should also be defined at this point. The kinds of sample-related information that can be important are summarized in Table 1.1.

Ideally, a complete description of the sample is available; for example, an aspirin tablet contains the active ingredient and various water-soluble excipients. The goal of HPLC separation in this case might be an assay of aspirin content, so the primary interest is in the chemical properties of aspirin. An-

1

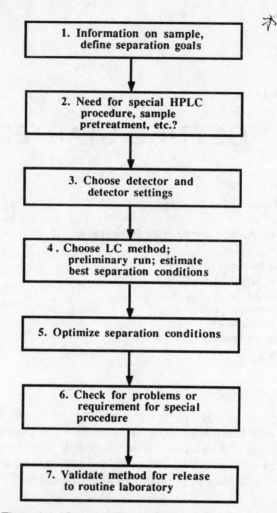

Figure 1.1   Steps in HPLC Method Development.

other situation might require analyzing a raw material for its major component and any contaminants. An example is provided by Fig. 1.2, which shows
possible components of crude samples of the pharmaceutical pafenolol (compound 6). In this case, the chemical structures of possible contaminants can
be inferred from the synthetic route used to prepare the compound, together
with known side reactions leading to by-products. A total of six compounds
can be expected in crude samples of pafenolol (compound 3 was ruled out
because of its instability).

TABLE 1.1   **Important Information Concerning Sample Composition and Properties**

- Number of compounds present
- Chemical structures (functionality) of compounds
- Molecular weights of compounds
- pK_a-values of compounds
- UV spectra of compounds
- Nature of sample matrix: solvent, fillers, etc.
- Concentration range of compounds in samples of interest
- Sample solubility

Many chromatographers rely on the chemical structure of sample compounds for clues to the best HPLC conditions. As discussed in later chapters, it is important to know if acids or bases are present, and what the pK_a-values are for different sample components. Likewise, it is important to know when high-molecular-weight compounds are present, or if chiral isomers are to be separated. The chromatographer may also be familiar with separations involving samples of similar structure. In these cases, it is tempting to try a similar procedure for the new sample. An alternative approach, emphasized in this book, is to *minimize the use of information on sample structure and carry out method development for most samples in the same general way.*

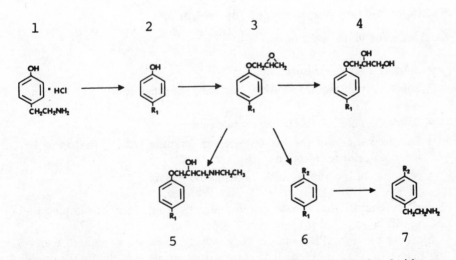

Figure 1.2   Compounds Present in Crude Samples of Pafenolol. Reprinted with permission from Ref. (1).

Separation Goals

The goals of HPLC separation need to be clearly specified. Some questions that should be asked at the beginning of method development include:

✗ • Is the primary goal analysis, or the recovery of sample fractions? (This book emphasizes analysis)
  • Is the chemical identity of all sample components known, or will qualitative analysis be required?
  • Is it necessary to resolve all sample components (e.g., enantiomers, diastereomers, homologs, oligomers, trace impurities)?
  • What precision is needed, if quantitative analysis is required?
  • For how many different sample matrices (raw materials, dosage forms, environmental samples, etc.) should the method be designed? Will more than one HPLC procedure be necessary?
  • Will the method be used for a few samples, or will a large number of samples be analyzed?
  • What HPLC equipment and skills are available in the routine laboratory that will use the final method?

Agreement on what is required of the method should be obtained before method development begins.

Nature of the Sample; the Need for Pretreatment

Samples come in different forms:

✗✓ • Solutions ready for injection
  • Solutions that require dilution, buffering, or other addition of a second liquid
  • Solids that are soluble in the mobile phase
  • Solutions that contain interferences or "column killers" that must be removed prior to injection
  • Mixtures of the analyte with an insoluble matrix

Except for samples that can be injected directly, some form of sample processing will be required prior to HPLC separation.

Most samples for HPLC analysis require weighing and/or volumetric dilution before they are injected. It is generally desirable that the final sample solution closely approximate the composition of the mobile phase. This suggests dissolving (or diluting) the sample in the mobile phase. Errors in volumetric dilution due to the use of volumetric flasks and pipettes generally must be taken into account, especially when the final assay result is required to be more precise than $\pm 1\%$ (coefficient of variation).

Many samples contain background interferences that greatly complicate HPLC separation. Other samples contain components that can damage the column, particularly for the repeated separation of similar samples. Generally, this kind of problem must be addressed before HPLC method development can begin. The usual approach is some kind of pre-HPLC separation of the sample (e.g., liquid- or solid-phase extraction). Alternatively, this kind of sample pretreatment can often be achieved on line with column switching.

Finally, some samples come dispersed in an insoluble matrix such as plant or animal tissue, highly cross-linked polymers, dried adhesives, etc. Here, some form of solvent extraction often will be required to dissolve the sample components of interest. Sample pretreatment is outside the scope of the present book, but for a good background of this subject, see Refs. (2-6).

Detection

Requirements for detectors vary with the separation goal. For measuring a single component, the ideal detector would sense only the material of interest and not respond to any other component. On the other hand, if qualitative analysis or preparative chromatography is the goal, universal detection is desired, so that every component in the mixture can be observed. For analysis, high detector sensitivity and a low minimum detection limit are desirable; for preparative isolations, high detector sensitivity seldom is essential.

UV photometric detectors should be used if possible, because of their excellent convenience and ruggedness. Therefore, it is useful to know something about the UV spectrum of each sample component before method development. This information, with a knowledge of the probable concentrations of various sample components, largely determines whether UV detection is feasible. Knowing the UV spectra of the various sample components (both analytes and interferences) also permits the choice of the best detection wavelength. Section 4.1 provides further information on when and how UV detection is best applied to borderline samples: those having marginal UV absorptivity and/or low analyte concentrations.

If a compound of interest does not possess an adequate UV chromophore, alternative means of detection should be considered, for example:

- Refractive index
- Fluorescence
- Electrochemical
- Precolumn or postcolumn derivatization

The newer differential refractometers have sensitivities of 10^{-8}–10^{-9} RI units and allow quantitation of as little as 100 ng of most compounds. These devices are essentially universal in their detectability but cannot be used with gradient elution. Compounds that fluoresce or are easily oxidized are ideal candidates for fluorescence or electrochemical detection, respectively. These detectors are capable of extreme sensitivity (femtomoles) for applicable compounds and can be used with gradient elution under some conditions. Finally, any compound having a suitable functional group can be derivatized to form a UV-absorbing or fluorescent product. Derivatization is usually a last choice, however, because of the added effort and decreased analytical precision. For further information on detectors and derivatization, see Refs. (2,3,7–14).

1.2 DEVELOPING THE SEPARATION

Method development for a sample should begin after the information, goals, and problems summarized in Sect. 1.1 are reviewed. The first attempt at separation requires selection of a promising set of experimental conditions. As reviewed in Table 1.2, several choices must be made. Many of these experimental conditions can be selected somewhat arbitrarily, based on the applicability of "standard" conditions for most samples. In some cases, use of conditions for previously successful separations of similar compounds is a good start. These initial conditions may not be optimum, but they will serve to get method development underway. In any case, method development can only start with an initial chromatogram.

Selecting an HPLC Method

For many samples, the initial choice of column and mobile-phase solvents can be guided by the recommendations of Table 1.2. However, there are some important exceptions and qualifications. Samples exhibiting any of the following characteristics will often require special consideration: (a) high-molecular-weight samples, such as synthetic polymers or proteins, (b) mixtures of optical isomers (enantiomers), (c) mixtures of other isomers, and (d) samples composed of inorganic salts or carbohydrates (sugars). For these special cases, we recommended that you start with Chapter 7, which provides a brief discussion of HPLC procedures for such samples.

Method development for most samples is often begun with reversed-phase HPLC. However, ion-pair and normal-phase chromatography are viable (perhaps preferred) options in many cases. Knowledge of the sample composition, together with possible experience with similar samples, may suggest a

TABLE 1.2 Experimental Conditions That Affect HPLC Separation

Separation Variable	Preferred Initial Choice
COLUMN	
Dimensions (length, i.d.)	25 × 0.46 cm
Particle size	5 μm
Stationary phase	C-8 or C-18
MOBILE PHASE	
Solvents A/B	Water/acetonitrile
%-B	Variable
Buffer (compound, pH, concentration)	25 mM phosphate, pH 3.5[a]
Additives (e.g., ion-pair reagents, amines)	See Table 9.5[a]
Flow rate	1–2 mL/min
TEMPERATURE	40°C
SAMPLE SIZE	
Volume[b]	≤ 50 μL
Mass[b]	≤ 100 μg

[a]These variables mainly affect the separation of ionized compounds.
[b]Assumes 25 × 0.46-cm reversed-phase column; values are smaller for smaller-volume columns and larger for larger-volume columns.

method other than reversed-phase. Table 1.3 reviews the characteristics and special advantages of each of these primary or "workhorse" HPLC methods. Table 1.4 is a similar presentation for some other (secondary) HPLC methods. The use of secondary HPLC methods is discussed further in Chapter 7.

Getting Started on Method Development

At this point, Step 4 of Fig. 1.1 should be considered: injecting the first sample. Unless there is a reason not to use reversed-phase HPLC for the sample, the next goal is to obtain a chromatogram using standard conditions (Table 1.2) that are likely to provide some resolution of the sample in this first run. One approach is to use an isocratic mobile phase of average "solvent strength" (intermediate percent organic). This strategy is likely to elute at least some compounds with reasonable retention, as illustrated by the separation of Fig. 1.3a (for a 10-component nitro-compound mixture). However, with some samples it may be necessary to wait quite a while for the last bands to clear the column before injecting the next sample.

TABLE 1.3   Characteristics of Primary HPLC Methods

Method/Description/Columns[a]	When Is the Method Preferred?
REVERSED-PHASE HPLC	
Uses water/organic mobile phase Columns: C-18 (ODS), C-8, phenyl, trimethylsilyl (TMS), cyano	First choice for neutral or nonionized compounds that dissolve in water/organic mixtures
ION-PAIR HPLC	
Uses water/organic mobile phase, a buffer to control pH, and an ion-pair reagent Columns: C-18, C-8, cyano	Good choice for ionic or ionizable compounds, especially bases or cations
NORMAL-PHASE HPLC	
Uses mixtures of organic solvents as mobile phase Columns: cyano, diol, amino, silica	Good second choice when reversed-phase or ion-pair HPLC are ineffective; first choice for lipophilic samples that do not dissolve well in water/organic mixtures; first choice for mixtures of isomers and for preparative-scale HPLC (silica best)

[a]All columns (except one) recommended here are packed with bonded-phase silica particles (see Chapt. 5 of Ref. 2). This list is representative but not exhaustive.

An alternative strategy is to use a stronger isocratic mobile phase (e.g., 100% organic) in the first run and then successively reduce the organic content in 20%-volume increments in succeeding runs. This technique is illustrated by the chromatograms of Figure 1.3b–d. In this case, 60%v methanol/water (Fig. 1.3d) gives a promising chromatogram in terms of desired k' range and analysis time, although not all 10 bands are adequately resolved.

A better approach is to use gradient elution for the initial sample separation, as illustrated in Fig. 1.3e. A gradient run can be designed to give some separation of the sample in the first run; this initial run can then guide further adjustments of the mobile phase composition for improved separation. Starting with gradient elution does not presuppose that a final method will use a gradient. Results from one or more gradient runs can be used to design analogous isocratic separations (see Chapters 2, 8, and 9).

Improving the Separation

Usually, the initial separation in some ways will be less than adequate. After a few additional tries, it may be tempting to accept a marginal separation.

TABLE 1.4   Characteristics of Secondary HPLC Methods

Method/Description/Columns	When Is the Method Preferred?
ION-EXCHANGE CHROMATOGRAPHY	
Uses aqueous mobile phase plus buffer for pH control	First choice for separating mixtures of inorganic ions (ion chromatography); good choice for separating protein and nucleic acid samples and related compounds
Columns: anion or cation exchange	
SIZE-EXCLUSION CHROMATOGRAPHY	
Uses either aqueous (gel filtration) or organic (gel permeation) mobile phases	Good first choice for separating high-molecular-weight samples such as proteins and synthetic polymers; used for molecular-weight-distribution measurements
Columns: diol-phase for gel filtration; polystyrene or silica for gel permeation; pore size dictates MW range	
HYDROPHOBIC-INTERACTION CHROMATOGRAPHY	
Uses salt solutions as mobile phase	Used for separating proteins
Columns: similar to reversed-phase packings but much less hydrophobic	

However, more experienced workers realize that a good separation, particularly for use as a routine procedure to analyze many samples, requires more than minimal resolution of the individual sample bands. Specifically, the experienced chromatographer will consider several aspects of the separation, as summarized in Table 1.5.

Separation or resolution is a primary requirement in a routine HPLC procedure. For samples containing five or fewer components, baseline resolution ($R_s > 1.5$) often can be easily attained for the bands of interest. This level of resolution favors maximum precision in reported results. Resolution usually degrades during the life of a column and can vary day to day with minor fluctuations in separation conditions. Therefore, values of $R_s \geq 2$ should be the goal during method development of simple mixtures. Samples containing 10 or more components generally will be more difficult to separate, and here the separation goal often should be relaxed to $R_s \geq 1$.

The time required for a separation (run time or retention time of last band) should be as short as possible. This assumes that the other goals of Table 1.5 have been achieved and that the total time spent on method development is

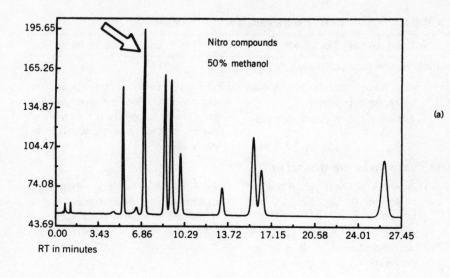

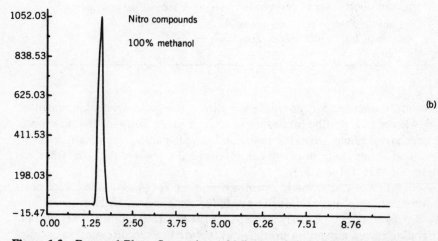

Figure 1.3 Reversed-Phase Separations of Mixture of Nitro-Aromatic Compounds. Column, 25 × 0.46-cm, 5-μm Zorbax C-8; flow rate, 2 mL/min; 35°C; methanol/water mixtures for mobile phase; sample components: nitrobenzene, benzene, 2,6-dinitrotoluene, 2-nitrotoluene, 4-nitrotoluene, 3-nitrotoluene, toluene, 2-nitro-1,3-xylene, 4-nitro-1,3-xylene, m-xylene. (a) 50% methanol/water; (b) 100% methanol/water; (c) 80% methanol/water; (d) 60% acetonitrile/water; (e) 5–100% methanol/water gradient in 15 min. Arrows indicate unresolved band-pairs.

10

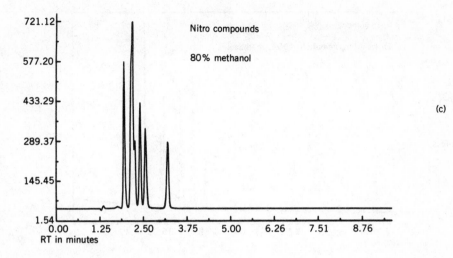

(c)

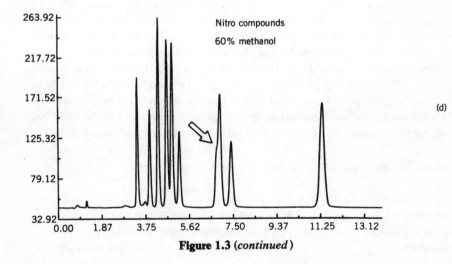

(d)

Figure 1.3 (*continued*)

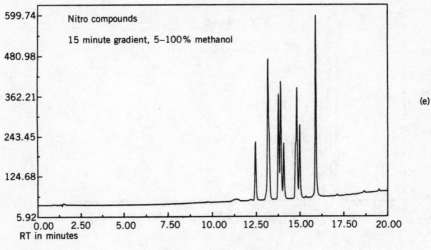

Figure 1.3 *(continued)*

reasonable. However, run time should be compared with the 2-hr setup time typically required for an HPLC procedure. Thus, if only two or three samples are to be assayed at one time, a run time of 20 min is not excessive. When 10 or more samples are to be assayed at the same time, then run times of 5–10 min are desirable. There is rarely any real reason to seek run times of a minute or less, although very fast separations are seldom detrimental.

The pressure required for the operating system has practical consequences. The operating pressure for most new columns should be less than 150 bar ($\sim$2000 psi). Since most HPLC pumps will operate at pressures of 300–400 bar, some workers will accept an operating pressure of 200–300 bar for a routine procedure. However, two considerations suggest a more conserv-

TABLE 1.5   Separation Goals in HPLC Method Development

Goal[a]	Comment
Resolution	Precise and rugged quantitative analysis requires $R_s > 1.5$
Separation time	<5–10 min is desirable for routine procedures
Quantitation	$\leq 2\%$ (1 std. dev.) for assays; 5% for less demanding or trace analyses
Pressure	<150 bar is desirable; <200 bar is usually essential (new column assumed)
Peak height	Narrow peaks for large signal/noise ratio are desirable
Solvent consumption	Minimum mobile-phase usage per run is desirable

[a]Roughly in order of decreasing importance, but may vary with analysis requirements.

ative target for maximum operating pressure during method development. First, during the life of a column, the back pressure may rise by a factor of up to twofold, due to the gradual plugging of the column by particulate matter. Second, at lower pressures (e.g., <150 bar), pumps, sample valves, and especially autosamplers, operate much better; seals last longer; columns tend to plug less; and system reliability is significantly greater. For these reasons, a target pressure of <50% of the maximum capability of the pump (typically, <100–150 bar) is desirable.

Improving an initial separation to meet the goals of Table 1.5 is the primary subject of this book.

Checking for Problems

As method development proceeds, various problems can arise. Initial chromatograms may contain bands that are noticeably broader than expected (lower column plate numbers), or bands may tail appreciably. Later, during use of the method, it may be found that replacing the original column with an "equivalent" column from the same manufacturer causes an unacceptable change in the separation. Consequently, a routine laboratory may not be able to reproduce the method on another (nominally equivalent) column. Column life also may be found to be undesirably short, e.g., less than 100 sample injections. For routine methods that are to be used for long time periods, it is important to anticipate and test for these and other problems before the method is released, rather than to try solving such problems later. Failure to deal with method problems at an early stage usually results in wasted time during method development and in subsequent routine analyses.

TABLE 1.6 Completing the Method[a]

1. Preliminary data to show required method performance
2. Written assay procedure developed for use by other operators
3. Systematic validation of method performance for more than one system/operator, using samples that cover the expected range in composition and analyte concentration; data obtained for day-to-day and interlaboratory operation
4. Data obtained on expected life of column and column-to-column reproducibility
5. Deviant results studied for possible correction of hidden problems
6. All variables (temperature, mobile-phase composition, etc.) studied for effect on separation; limits defined for these variables; remedies suggested for possible problems (poor resolution of key band pair, increased retention for last band with longer run times, etc.)

[a]Primarily applicable to routine or quality-control methods.

∠

Completing the Method

The final procedure should meet all the goals that were defined at the beginning of method development. The method should also be robust in routine operation and usable by all the laboratories for which it is intended. Achieving all of these objectives requires attention to the various items of Table 1.6.

The preceding discussion applies to HPLC methods that must meet stringent standards of precision, accuracy, ruggedness, and transferability. In other cases, all that might be required is a quick, "rough" answer to a specific problem. For these samples, many of the requirements of Tables 1.5 and 1.6 can be relaxed or eliminated. Some of the steps of Fig. 1.1 may also prove unnecessary. Common sense and an awareness of the actual goals of each method-development assignment are then required.

REFERENCES

1. S. O. Jansson and S. Johansson, *J. Chromatogr., 242* (1982) 41.

2. L. R. Snyder and J. J. Kirkland, *An Introduction to Modern Liquid Chromatography,* 2nd ed., Wiley-Interscience, New York, 1979, Chapts. 16, 17.

3. H. Engelhardt, ed., *Practice of High Performance Liquid Chromatography,* Springer-Verlag, Berlin, 1986, pp. 143–178.

4. I. W. Wainer, ed., *Liquid Chromatography in Pharmaceutical Development: An Introduction,* Aster, Springfield, Ore., 1985, pp. 133–148, 323–376.

5. P. Dimson, S. Brocato, and R. E. Majors, *Am. Lab. (Fairfield, Conn.), 18* (1986) 82.

6. K. Ramsteiner, *Int. J. Environ. Anal. Chem., 25* (1986) 49.

7. J. F. Lawrence and R. W. Frei, *Chemical Derivatization in Liquid Chromatography,* Elsevier, Amsterdam, 1976.

8. S. Ahuja, *J. Chromatogr. Sci., 17* (1979) 168.

9. K. Blau and G. S. King, eds., *Handbook of Derivatives for Chromatography,* Heyden, London, 1978.

10. D. R. Knapp, *Handbook of Analytical Derivatization Reactions,* Wiley, New York, 1979.

11. R. W. Frei and J. F. Lawrence, eds., *Chemical Derivatization in Analytical Chemistry, Vol. 1,* Plenum Press, New York, 1981.

12. C. F. Poole and S. A. Schuette, *Contemporary Practice of Chromatography,* Elsevier, Amsterdam, 1984, pp. 511–521.

13. R. P. W. Scott, *Liquid Chromatography Detectors,* 2nd ed., Elsevier, Amsterdam, 1985.

14. J. F. Lawrence, ed., *J. Chromatogr. Sci., 17* (1979) 115–172.

2

BASICS OF SEPARATION: MOBILE-PHASE EFFECTS

Most chromatographers have some idea of how changing experimental conditions affects an HPLC separation. They know that increasing the mobile-phase percent organic in a reversed-phase run will elute the bands sooner, which usually decreases resolution. Or, if the flow rate is decreased, the run time increases but the separation usually gets better. Chromatographers also know that trying a different column often changes the separation, sometimes for the better. This knowledge of how conditions affect separation is usually a combination of training and experience. But, often what is known is a mixture of "soft" and "hard" facts. This chapter and the next review some basics of HPLC separation: "hard facts" that will make sure that method development starts out in the right direction.

An important step in method development is to use some of the more important information that has been uncovered in the past few years. At the present time, HPLC publications (articles, reviews, books) are appearing in unprecedented numbers, and this information explosion is overwhelming most workers. This book summarizes and organizes those findings that really make an impact on method development.

Finally, although the basic theory of HPLC has been under development for several decades, most workers use this information only in a qualitative sense. They decide whether an experimental variable should be increased or decreased, to improve separation or shorten run time. But the perception is that this basic theory does not tell *how much* to change a particular variable, or *which variable* is most critical. One of the major developments of the past half-dozen years is the use of basic concepts to make quantitative predictions of how the separation will change when some variable is altered by a given amount. These basic concepts can also predict the change in separation when several conditions are varied simultaneously. To make the most of this new opportunity, however, we will usually need a computer. Chapter 8 provides a detailed examination of this area.

2.1   RESOLUTION: GENERAL CONSIDERATIONS

The quality of a separation is measured by *resolution,* the degree of disengagement of two bands. Bands that overlap badly have small values of the resolution function, R_s:

$$R_s = (t_2 - t_1)/[(1/2)(W_1 + W_2)]. \tag{2.1}$$

Here, t_1 and t_2 refer to the retention times of two adjacent bands, and W_1 and W_2 are their baseline bandwidths. R_s is referred to as the resolution of bands 1 and 2. Reference 1 provides a thorough discussion of how resolution is measured, what minimum resolution is required for an HPLC separation, and how resolution changes with changes in separation conditions. In this chapter, we will summarize the most important parts of this discussion and add some new material to enhance its usefulness.

Measurement of Resolution

Equation 2.1 is a convenient way of measuring R_s when bands 1 and 2 are completely separated ($R_s \geq 1.5$). However, baseline bandwidths are not easy to measure accurately when peaks overlap, so a better approach is to use the corresponding bandwidths at half-height, w_1 and w_2:

$$R_s = 1.18 \, (t_2 - t_1)/(w_1 + w_2). \tag{2.2}$$

Other equations based on different bandwidths can be derived from the usual relationships for a Gaussian curve (Fig. 2.1).

When resolution is poor ($R_s < 0.7$), or when only a qualitative estimate of R_s is required, standard resolution curves as in Figs. 2.2–2.7 can be used to estimate R_s. The actual band-pair of interest is compared with the examples of Figs. 2.2–2.7 to find the best match. The resolution of the best-fit standard resolution curve is then assumed equal to R_s for the actual band-pair of interest. The examples of Figs. 2.2–2.7 provide additional information that can be of interest in method development: interference of one band with the other for peak-height or peak-area measurement, purity of separated bands, or change in retention time because of band overlap (as described more fully in Ref. 1).

Alternatively, a convenient and more accurate measurement of R_s is possible for bands that overlap moderately. This approach is based on the height of the valley between the two bands (Fig. 2.8). First, the height of the valley from the baseline is measured (20 mm for this example). Then, the valley-height is expressed as a percentage (74%) of the height of the smaller band (27 mm). The peak-height ratio of the two bands must also be measured (54 mm/27 mm = 2/1); Table 2.1 then gives an R_s value of 0.80 for the example shown in Fig. 2.8 (interpolation between 70% and 80% for a peak-height ratio of 2/1).

Whatever approach is adopted for measuring R_s, it is important during method development to have a quantitative measure of the resolution of those bands of interest that are least separated in each chromatogram.

Minimum Resolution

Chromatograms that contain more than two bands will have different R_s-values for each band-pair. Usually the emphasis is only on that band-pair(s) that

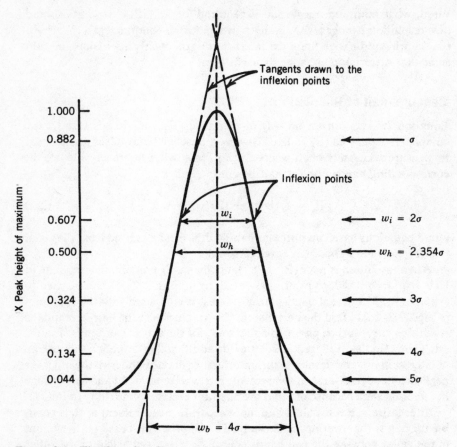

Figure 2.1   The Gaussian Curve and the Interrelationship of Different Bandwidth Measurements. Reprinted with permission from Ref. (2).

(a) has the smallest value of R_s and (b) includes a band that must be analyzed or separated. This consideration is illustrated by Fig. 2.9a, for the separation of a mixture of nitro-aromatic compounds and their hydrocarbon precursors. In this case, all 10 major bands are of interest. The poorest-resolved band-pair can quickly be identified by examining the valley between each band-pair. Bands 4 and 5 appear to be the poorest resolved, with bands 2 and 3 a close second. Use of Table 2.1 provides R_s-values for each band-pair: 1.24 (band-pair 4,5) and 1.32 (band-pair 2,3). Therefore, the R_s value for this separation (Fig. 2.9a) is 1.24. If greater resolution is required for this sample, R_s for bands 4 and 5 will have to be increased.

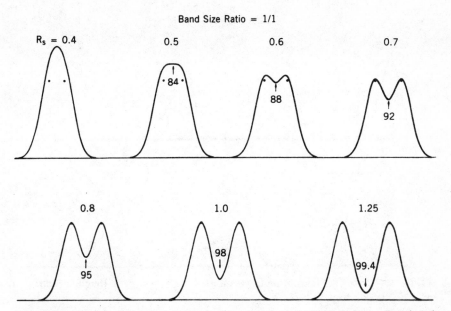

Figure 2.2 Standard Resolution Curves for a 1/1 Ratio of Peak Heights. Reprinted with permission from Ref. (1).

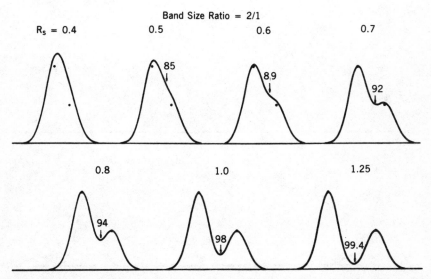

Figure 2.3 Standard Resolution Curves for a 2/1 Ratio of Peak Heights. Reprinted with permission from Ref. (1).

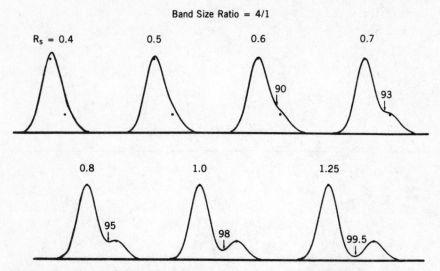

Figure 2.4   Standard Resolution Curves for a 4/1 Ratio of Peak Heights. Reprinted with permission from Ref. (1).

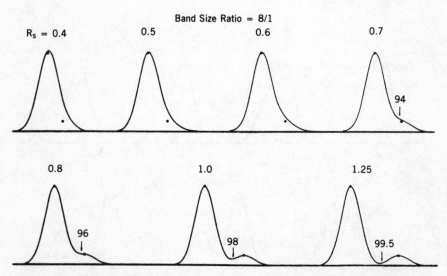

Figure 2.5   Standard Resolution Curves for a 8/1 Ratio of Peak Heights. Reprinted with permission from Ref. (1).

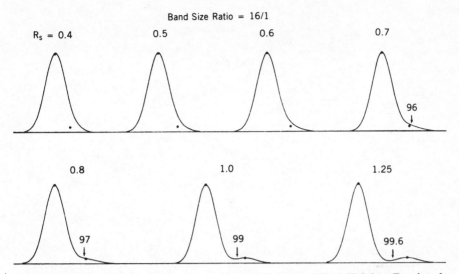

Figure 2.6 Standard Resolution Curves for a 16/1 Ratio of Peak Heights. Reprinted with permission from Ref. (1).

A change in some experimental variable usually causes a change in band resolution. In many cases, the critical band-pair that gives the lowest R_s-value for one separation condition will not be the same for some other condition. This effect is illustrated in Fig. 2.9b for the same sample, but with the mobile phase changed from 55% methanol to 60% methanol. Now, bands 7 and 8 overlap most severely, with R_s equal to about 0.6 (estimated using Fig. 2.3). In situations such as this (percent methanol changed but other conditions the same), it is often useful to plot resolution as a function of the condition being varied, as illustrated in Fig. 2.10 for the system of Fig. 2.9. Here, a *relative resolution map* (RRM) is shown for this 10-component sample as a function of percent methanol in the mobile phase. An (RRM) is a plot of the resolution for the poorest-resolved band-pair as a function of percent organic in the mobile phase, assuming a constant plate number (N) for all peaks (N = 10,000 in Fig. 2.10). The RRM of Fig. 2.10 indicates that band crossovers occur for several mobile-phase compositions where R_s goes to zero. Such a map shows which percent methanol will give the best separation of the sample, that is, the chromatogram with the largest value of R_s for the most overlapped band-pair in the sample. The map of Fig. 2.10 indicates that 55% methanol is close to the optimum value. Later, we will discuss various ways of mapping resolution as a function of separation conditions.

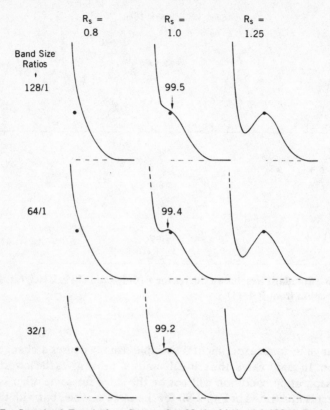

Figure 2.7 Standard Resolution Curves for 32/1, 64/1 and 128/1 Ratios of Peak Heights. Reprinted with permission from Ref. (1).

The resolution of the poorest-resolved band-pair is not the only significant measure of resolution. Usually, the resolution of other band-pairs will be less interesting, but consider the hypothetical examples of Figs. 2.11a and b. The minimum R_s value (for bands 1 and 2) is the same in both cases, but the separation of Fig. 2.11a is superior. The reason is that, for the same minimum resolution, run time is shorter. Or, consider the examples of Figs. 2.11c and d. In this case, the number of compounds in the sample is unknown. The apparent resolution of the most overlapped band-pair is better in Fig. 2.11c than in 2.11d, but there are more bands in 2.11c! Clearly, one or more band-pairs is unresolved in Fig. 2.11d. In this case, the resolution of Fig. 2.11c is better ($R_s = 0$ for Fig. 2.11d). For reasons that should be clear from these examples, several workers have proposed additional criteria to quantitatively evaluate separation (5). However, we feel that these alternative measures of

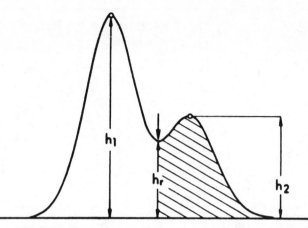

Figure 2.8   Measurement of Height of the Valley between Two Overlapping Bands. Reprinted with permission from Ref. (1).

TABLE 2.1   Estimating Resolution from the Height of the Valley between Two Adjacent Bands

h_r (%)	R_s for Indicated Ratio of Band Size			
	1/1	2/1	4/1	8/1
10	1.22	1.26	1.30	
20	1.07	1.13	1.17	1.31
30	0.97	1.05	1.10	1.22
40	0.89	0.99	1.05	1.13
50	0.83	0.92	1.01	1.07
60	0.78	0.86	0.96	1.01
70	0.74	0.82	0.91	0.96
80	0.70	0.77	0.86	0.92
90	0.66	0.74	0.82	0.89

Source: Reprinted with permission from Ref. (4).

expressing separation quality are generally too cumbersome to be useful. A value of minimum R_s plus common sense (count the peaks!) will be the best approach in most cases (see discussion of Ref. 3).

How Much Resolution is Required?

For quantitative analysis, baseline resolution ($R_s \geq 1.5$) of all bands of interest is strongly recommended. Baseline separation simplifies the measurement

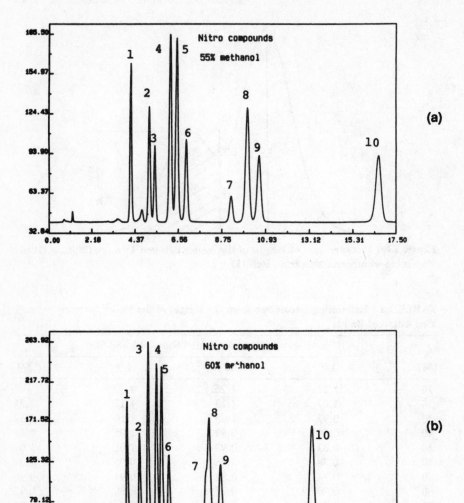

Figure 2.9 Separation of Nitro-aromatic Sample shown in Fig. 1.3. Column, 25 × 0.46-cm, 5-μm Zorbax C-8; flow rate, 2 mL/min; (a), 55% methanol/water; (b), 60% methanol/water. Reprinted with permission from Ref. (6).

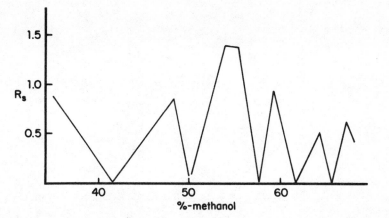

Figure 2.10   Resolution Map for Separation of Fig. 2.9 as a Function of Percent Methanol. Assumes a 10,000-plate column. Reprinted with permission from Ref. (6).

of band-area or peak-height values, because the position of the baseline near the peaks is easier to determine. Resolution in excess of $R_s = 1.5$ also results in a more rugged separation. During the use of the final HPLC procedure in different laboratories, some variability in the separation will inevitably be encountered. Excess resolution in the starting method then provides some assurance that small changes in the separation will not compromise the method.

As a column is used, its performance will inevitably degrade. This results in broadening of sample bands, often accompanied by a decrease in separation factors and a decrease in resolution. Band-pairs with $R_s \geq 2$ for a new column are more likely to be suitably resolved after several hundred analyses on the same column. That is, larger values of R_s in the first chromatogram with a new column generally mean that the column can be used longer with that method.

Some samples, such as that of Fig. 2.9, are sufficiently complex that it may be difficult to achieve $R_s \geq 2$ for every band-pair. Either there is not enough room in the chromatogram to allow this much separation of every band, or some bands will have low α-values under all conditions. It may be necessary to accept smaller values of R_s in some cases, but it is then particularly important to maximize R_s as much as possible. Quantitation with $R_s < 1$ is generally less precise, particularly since R_s will vary from day to day and from lab to lab.

Separation varies with each of the conditions listed in Table 1.2, as well as with other factors. Some of these variables affect resolution in a major way,

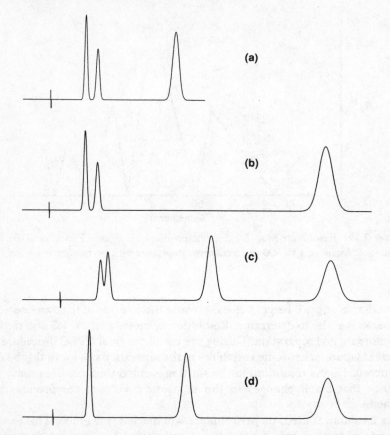

Figure 2.11 Hypothetical Examples Showing Limitations of R_s as a Measure of Resolution for a Chromatogram (see text). (a,b), same sample; (c,d), different sample.

while other separation parameters are relatively unimportant. To understand how to control resolution by choosing the best set of experimental conditions, it is necessary to begin with the basic equation for resolution:

$$R_s = (1/4)(\alpha - 1) N^{0.5} [k'/(1 + k')]. \qquad (2.3)$$

Here, α is the separation factor, equal to k_2/k_1; k_1 and k_2 refer to the capacity factors (k') for bands 1 and 2 of a given band-pair; N is the column plate number (average for the two bands); k' refers to the average k'-value for the two bands. When N is large and the two bands are barely resolved (the usual case of interest), the k'-value for either band in Eqn. 2.3 can be used. A value of N for any band is conveniently measured from the retention time t_R, and bandwidth at half-height, $w_{1/2}$:

$$N = 5.56(t_R/w_{1/2})^{1/2} \qquad (2.3a)$$

According to Eqn. 2.3, resolution is determined by three separate factors: the separation factor α, the plate number N, and the capacity factor k'. These three parameters are more or less independent of each other, so that conditions can be varied to optimize first k', then α, and finally N. While this is true to a first approximation, actually all three parameters are interrelated. Thus, if conditions are changed so as to vary k', significant changes may also occur in both α and N. Nevertheless, it is good strategy to first adjust conditions for an optimum k'-value and then focus attention on conditions that affect α and N. In this approach, k'-values for the sample are held roughly constant, while conditions are changed for improved values of α and N.

2.2 RESOLUTION VS. CONDITIONS: SOLVENT STRENGTH (k')

Resolution as a Function of k'

Equation 2.3 states that R_s increases as $k'/(1 + k')$ increases. That is, larger k'-values lead to a continuing increase in resolution. Unfortunately, an increase in k' also leads to longer run times and to wider bands, which are harder to quantitate accurately. So, intermediate values of k' are desirable, in the range of $1 < k' < 20$ for all bands in the chromatogram. Even when the resolution of bands with $k' < 1$ is possible, the resulting separation will usually be unsatisfactory: interfering bands are more likely in the region $0 < k' < 1$, and poor baselines that make quantitation difficult often are observed in this part of the chromatogram. For this reason, every effort should be made to avoid conditions that lead to bands with $k' < 1$.

The k'-values of individual bands increase or decrease with changes in solvent strength. For the case of reversed-phase HPLC, solvent strength increases with the percent organic in the water/organic mobile phase. Typically, an increase in percent organic by 10%v will decrease k' for every band by a factor of 2 to 3. An example is shown in Fig. 2.12 for the same nitro-aromatic mixture of Fig. 2.9, with methanol varied from 30 to 70%v. The 30%-methanol separation (Fig. 2.9a) has a run time of 3 h, but all 10 compounds are well resolved. The 70%-methanol run (Fig. 2.12e) has a run time of only 6 min, but only eight peaks can be seen; resolution of other bands is marginal. The actual height of the last band is 40-fold greater in the 70%-

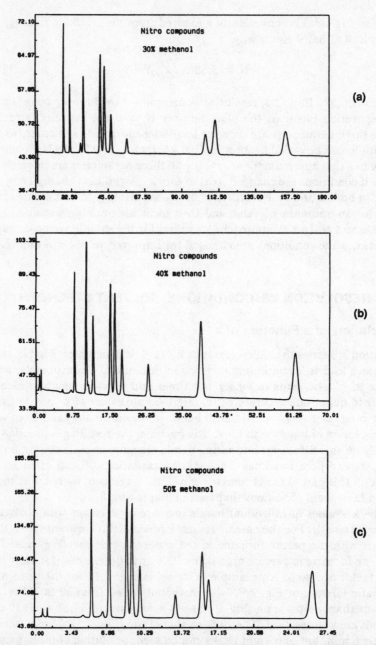

Figure 2.12 Separation of Nitro-aromatic Sample Shown in Fig. 2.9, Similar Conditions. Mobile phase varied as indicated. Reprinted with permission from Ref. (6).

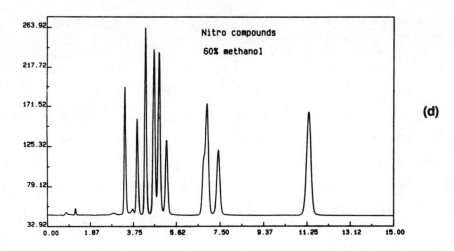

(d)

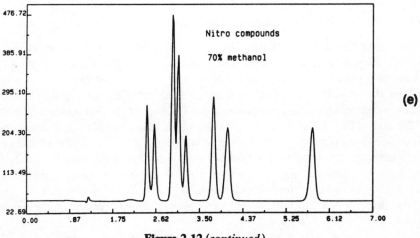

(e)

Figure 2.12 (*continued*)

methanol run compared with the 30%-methanol run. (Different detector attenuations were used to obtain these figures.)

It is useful to measure the k' range of the sample as early as possible in method development. This range is expressed as the ratio $(k'$ for last band)/$(k'$ for first band) $= R_k$. For example, the k' range for the sample of Fig. 2.12d $(3 < k' < 11)$ corresponds to $R_k = 4$. The R_k of a sample has a significant impact on the approach to method development. Thus, for different values of R_k, we recommend the following:

$R_k > 20$ Use gradient elution (see Chapter 6).

$R_k = 20$ Select a solvent strength that gives a chromatogram with a range of $1 < k' < 20$.

$R_k \lll 20$ Select a solvent strength for minimum k' values; i.e., $k' = 1$ for the first band, or vary solvent strength for control of α.

With $R_k \lll 20$, the rapid completion of method development is most likely. Fortunately, most samples (except those of biological origin) fit into this category. The possibility of using a change in solvent strength to vary α is discussed in Section 2.3.

Determining Optimum Solvent Strength

Conditions that produce the preferred solvent strength are best determined empirically. One approach is to begin with a mobile phase that is probably too strong and reduce solvent strength to increase k' between successive runs. When all the bands fit within the range of $1 < k' < 20$, the mobile phase is near optimum from the standpoint of solvent strength. This procedure is illustrated in Fig. 2.12 for the nitro-aromatic mixture; mobile phases containing 50-70%v methanol/water meet the criterion $1 < k' < 20$ for all bands.

Another approach is to use gradient elution for the initial run. Figure 2.13 shows a gradient run for a sample of the same nitro-aromatic compounds discussed previously. From this experiment, it is possible to estimate the approximate solvent composition that will give sample k'-values in the range $1 < k' < 20$ (see Sect. 9.3).

First, determine the time during the gradient when the midpoint of peak elution, t_x, occurs (halfway between the first and last band eluted) (see Fig. 9.6). In Fig. 2.13, t_x occurs at about 14 min. Next, use the following equation to determine the time t_k during the gradient (5-100% methanol/water in 15 min) when the appropriate mobile phase leaves the gradient mixer:

$$t_k = t_x - 2 t_0 - t_D. \tag{2.4}$$

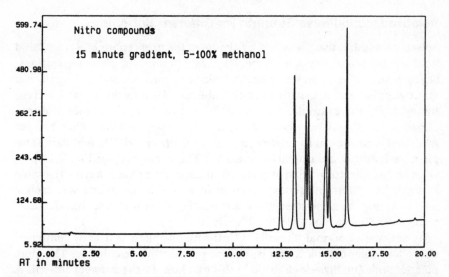

Figure 2.13 Separation of Nitro-aromatic Sample Shown in Fig. 2.9 by Gradient Elution. Similar conditions, except 5–100% methanol/water gradient, 15-min gradient time. Reproduced from the *J. Chromatogr. Sci.* (Ref. 6) by permission of Preston Publications.

Here t_0 is the column dead time, equal to 1.2 min for this separation. The system dwell time t_D is the time it takes for solvent to flow from the gradient mixer through the pump and associated tubing to the column as described in Chapter 6; here, $t_D = 2.8$ min. Equation 2.4 then yields $t_k = 14 - 2(1.2) - 2.8 = 8.8$ min. The gradient goes from 5 to 100% methanol in 15 min ($t_G = 15$ min), so that at 8.8 min the methanol concentration will be:

$$\% \text{ methanol} = (\text{initial }\%) + (t_k/t_G)[(\text{final }\%) - (\text{initial }\%)]$$
$$= (5 + 95[8.8/15])\%$$
$$= 61\% \text{ methanol.}$$

This mobile-phase composition (61%v methanol/water) should give acceptable k'-values for this sample. Using 61% methanol/water as the mobile phase, the k' range for the sample of Fig. 2.13 was measured to be $1.6 < k' < 9$, that is, in the middle of the desired range $1 < k' < 20$.

An exact approach for determining the optimum solvent strength is to use two gradients but with different gradient times for each run. This technique allows the k' for each band to be calculated as a function of any mobile phase composition (e.g., percent organic). This procedure is described in Chapters 6 and 8.

Relative Mobile-Phase Strength for Different Solvents

Solvent strength is usually varied by changing the proportions of a strong and weak solvent in a binary-solvent mobile phase. In reversed-phase separations, organics such as methanol and acetonitrile are strong and water is weak. For selectivity changes, it sometimes is desirable to substitute one organic solvent for another, for example, methanol for acetonitrile, or tetrahydrofuran (THF) for methanol. In these cases, it is necessary to know what percent methanol is equivalent to what percent acetonitrile or THF, to maintain sample k'-values that are roughly equivalent. The nomograph of Fig. 2.14 provides for the interconversion of reversed-phase mobile phases having the same strength (iso-eluotropic series). Vertical lines in this figure intersect mobile phases having the same strength. For example, 50%v methanol has the same strength as 40% acetonitrile or 30% THF.

In the case of normal-phase HPLC, the mobile phase is usually composed of two less-polar organics, that is, hexane plus methylene chloride. The more polar organic (methylene chloride in this case) will be the stronger solvent. A twofold decrease in the percent strong organic (e.g., methylene chloride) will generally result in an increase in k' for all bands by a factor of about 3. Thus, if the mobile phase initially is 50% methylene chloride/hexane and is changed to 25% methylene chloride/hexane, it can be expected that sample k'-values will increase by about threefold for each band. Table 2.2 shows solvent strength values for some useful organic solvent mixtures, similar to that of Figure 2.14 for reversed-phase HPLC.

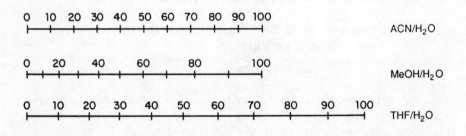

MeOH = methanol, ACN = Acetonitrile, THF = tetrahydrofuran

Figure 2.14   Solvent-Strength Nomograph for Reversed-Phase HPLC. Adapted from data of Ref. (9,10).

**TABLE 2.2   Equal-Strength Mobile Phases for
Liquid-Solid Chromatography (Silica)**

	Percent by Volume Mixture for Indicated Solvent Strength (ϵ°-Value^a)						
ϵ°	MC/hex	MTBE/hex	ACN/hex^b	EA/hex	MTBE/MC	ACN/MC	MeOH/MC
0.05	3.5			—			
0.10	10	0.2	0.3	0.3			
0.15	18	0.6	0.6	0.8			
0.20	32	1.4	1.1	1.8			
0.25	58	4.3	2	6			
0.30	100	13	3.5	18			
0.35		35	8	35	30	12	
0.40		57	24	52	60	30	3.5
0.45		84	52	75	88	55	6
0.50			88			88	9
0.60							16
0.70							28
0.80							52
0.90							95

^{a}MC, methylene chloride; hex, hexane or FC-113; MTBE, methyl-t-butyl ether; ACN, acetonitrile; EA, ethyl acetate; MeOH, methanol.
^{b}Acetonitrile and hexane are not miscible; use FC-113 instead of hexane.
Source: Reprinted with permission from Ref. (15).

2.3   RESOLUTION VS. CONDITIONS: MOBILE-PHASE SELECTIVITY (α)

Once the isocratic solvent strength has been properly adjusted for the sample, the next separation parameter that can be explored is α. In many cases, it is possible to obtain a satisfactory separation simply by varying solvent strength. This will usually be true for simple, easily resolved samples; that is, a sample containing only two or three components. However, even samples as complex as the nitro-aromatic mixture (Figs. 2.9 and 2.12) may be adequately resolved through changes in solvent strength alone. In such cases, the k' range of the sample (R_k) must be $\ll 20$, so that significantly different solvent strengths (percent organic values) are available with $1 < k' < 20$. For the nitro-aromatic sample, mobile phases of 50–70% methanol qualify. The possibilities of varying solvent strength are shown in the examples of Figs. 2.9a and 2.12. Thus, all 10 bands can be separated with either 55% methanol as mobile phase, or with 30% methanol (but note the differences in run time); overlapping bands are observed in the other examples of Figs. 2.9 and 2.12. The variation of percent organic as a simple means for optimizing band spacing appears to be broadly applicable (7,8). The use of solvent strength for varying α and band spacing is discussed further in Chapter 4.

Apart from changes in solvent strength, several other separation variables can be altered to change band spacing and α. Some of the more commonly used variables are listed in Table 2.3 with a brief comment on each. The relative power of each variable in Table 2.3 to affect α generally decreases from first (organic solvent) to last (temperature).

Choice of Organic Solvent, Mobile-Phase pH, and Mobile-Phase Additives

Mobile-phase composition has a major effect on band spacing; no other variable will generally prove as powerful for controlling values of α. A trial-and-error approach to the selection of the best strong solvent, pH, and additives is

TABLE 2.3   Separation Conditions Used to Vary Band Spacing (α)

Variable	Comment
Choice of organic solvent	A change from methanol to acetonitrile or THF in reversed-phase HPLC often results in large changes in α; similar changes in α result in normal-phase HPLC for changes among methylene chloride, methyl-t-butyl ether (MTBE) and either ethyl acetate or acetonitrile.
Mobile-phase pH	A change in pH can have a major effect on band spacing for samples that contain acidic or basic compounds.
Solvent strength	A change in percent organic often provides significant changes in α for reversed-phase, ion-pair, or normal-phase HPLC; easiest way to optimize band spacing.
Concentration of mobile phase additives	The most common additive for varying α-values is an ion-pair agent; other additives include amine modifiers, buffers, and salts (including complexing agents).
Column type	This refers to the choice of bonded-phase for reversed-phase HPLC (C-18, phenyl, cyano, trimethyl, etc.), or to silica, alumina and various polar bonded-phase columns for normal-phase HPLC.
Temperature	The temperature can usually be varied between 0 and 70° C for purposes of controlling α-values; however, temperatures of 25–60° C are more common; temperature variation is not useful for normal-phase HPLC.

possible for a particular sample, but is time consuming and inefficient. It is better to try a minimum number of mobile phases that are most likely to provide maximum changes in α. Then, mobile phases can be blended, if necessary, for an optimum spacing of all the sample bands. This approach is illustrated in the example of Fig. 2.15, for the separation of a six-component sample by reversed-phase HPLC. A water/methanol mobile phase was first adjusted to give optimum solvent strength: 50% methanol in this example (Fig. 2.15a). However, the first two bands (1 and 2) are poorly resolved, so a change in band spacing was sought by substituting the strong solvent tetrahydrofuran (THF) for methanol (Fig. 2.15b). (Note that the percent organic in the mobile phase was changed when substituting THF [32%] for methanol [50%]; see the nomograph of Fig. 2.14). Use of THF resulted in an excellent resolution of bands 1 and 2, but now bands 2 and 3 overlap because retention for band 2 has been increased too much. These results suggest that some blend of 50% methanol and 30% THF will provide a good separation of all three bands. This is actually the case for a 2/1 blend of these two mobile phases (Fig. 2.15c, 25% THF plus 17% methanol).

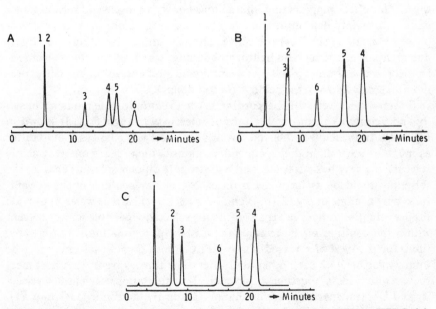

Figure 2.15 Effect of Strong Solvent on Band Spacing in Reversed-Phase HPLC. (a) 50% methanol/water; (b) 32% THF/water; (c) 25%-THF/17%-methanol/water. Bands are: 1, benzyl alcohol; 2, phenol; 3, 3-phenyl propanol; 4, 2,4-dimethyl phenol; 5, benzene; 6, diethylphthalate. Reprinted with permission from Ref. (10).

Chapter 4 presents a systematic approach for finding a mobile phase (based on different solvents, additives, or pH) that can provide optimum band spacing. In this section only the basis of such optimization procedures is discussed. What is needed now is a definition of the minimum number of unique mobile phases that can each provide maximum changes in α. Such an approach limits the number of different mobile-phase compositions that must be tried during method development, and provides some assurance that the best mobile-phase composition will be found. The selection of such a solvent set requires some knowledge about the retention process and the role played by individual solvents. Since retention is basically different in reversed-phase, ion-pairing and normal-phase systems, a separate discussion of each form of HPLC is required.

Reversed-Phase HPLC

Reversed-phase systems are characterized by strong interactions between the polar mobile phase and various sample molecules. On the other hand, interactions between sample molecules and the nonpolar stationary phase are weak. This effect suggests that interactions between sample and solvent molecules will mainly determine relative retention and values of α in reversed-phase separation (11). These selective solvent-sample interactions are largely due to dipole attraction and hydrogen bonding, which means that solvent selectivity can be characterized by solvent dipole moment, solvent basicity (proton acceptor), and solvent acidity (proton donor).

Solvents can be classified according to their *relative* dipole moment, basicity and acidity by means of the solvent-selectivity triangle (12,13) shown in Fig. 2.16. In principle, preferred solvents for selectivity changes should differ as much as possible in their polar interactions, which means one very acidic solvent, one very basic solvent, and one strongly dipolar solvent. That is, the solvents should be as far apart from each other as possible in the solvent-selectivity triangle of Fig. 2.16. Blends of such solvents, plus water to provide proper retention range, can then mimic the selectivity possible for any solvent within the confines of the triangle (as a first approximation). At the same time, for the case of reversed-phase HPLC, these organic solvents must be totally miscible with each other and water. The three solvents that best meet these requirements, together with practical considerations such as low viscosity and UV transparency, are methanol (Group II), acetonitrile (Group VI), and tetrahydrofuran (Group III). Four-solvent mobile-phase optimization using these three organic solvents plus water provides significant control over α-values in reversed-phase HPLC (14).

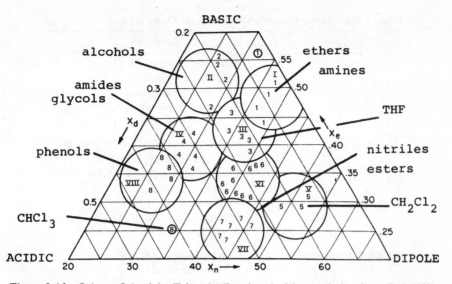

Figure 2.16   Solvent-Selectivity Triangle. Reprinted with permission from Ref. (13).

Normal-Phase HPLC

When the mobile phase is less polar and the stationary phase is more polar, this form of HPLC is called normal-phase chromatography. In this case, the stationary phase is sometimes referred to as an *adsorbent*. (Another term often used for normal-phase HPLC is *adsorption chromatography*.) Column packings commonly used in HPLC include silica and various polar-bonded phases (cyano, diol, amino). Unlike the case of reversed-phase chromatography, sample–solvent interactions are now relatively weak, and sample–adsorbent or solvent–adsorbent interactions are strong (15). This change in the relative importance of sample–solvent interactions (compared with reversed-phase HPLC) leads to a different classification of solvent selectivity (16,17).

In normal-phase chromatography, solvents can be classified (according to selectivity or the ability to affect α) into the four groups of Table 2.4. For less polar samples, where normal-phase HPLC is generally favored, the first three solvent groups of Table 2.4 (excluding proton donors) are preferred. Mobile phases for maximum control over band spacing will include a total of four solvents: one solvent from each of the first three groups of Table 2.4 plus a weak solvent to control solvent strength independently of selectivity. Hexane or (better) 1,1,2-trifluoro,1,2,2-trichloroethane (Freon® FC-113) is generally preferred as the weak solvent. Useful solvent combinations of this type, which also possess UV transparency for easy detection, include (a) the set of methy-

TABLE 2.4 Solvent Selectivity for Normal-Phase HPLC

Selectivity Type	Solvents[a]
Nonlocalizing	Methylene chloride (0.30)
	Chloroform (0.26)
	Chloropropane (0.28)
	Aromatic hydrocarbons (0.20–0.25)
	Carbon tetrachloride (0.11)
Basic localizing	Methyl-t-butyl ether (MTBE) (0.48)
	Tetrahydrofuran (THF) (0.53)
	Ethyl or isopropyl ether (0.43)
	Dioxane (0.51)
	Pyridine (>0.6)
	Dimethyl sulfoxide (>0.6)
	Alkyl amines (>0.6)
Nonbasic localizing	Ethyl acetate (0.48)
	Acetonitrile (0.52)
	Acetone (0.53)
	Nitromethane (0.5)
Proton donors	Alcohols (>0.6)

[a]Relative solvent strength values ($\epsilon°$) in parentheses.
Source: Adapted with permission from Ref. (15).

lene chloride, methyl-t-butyl ether (MTBE), and acetonitrile in FC-113 and (b) the set of methylene chloride, MTBE (or other ether), and ethyl acetate in hexane. Either of these two mobile-phase combinations is miscible over the entire range of composition.

Figure 2.17a illustrates the variation in k′ (and α) that can occur for a typical sample separated by normal-phase HPLC, when the proportions of three selective solvents in the mobile phase are varied. Figure 2.17b shows the optimum separation of this sample using a three-solvent mixture. None of the binary solvents in this case could provide resolution of all 13 compounds.

Ion-Pair HPLC

The preceding discussion of reversed-phase HPLC assumes nonionic sample compounds. Ion-pair chromatography is used when some or all of the sample components are ionized and, therefore, can interact with an ion-pair agent (19). This means that retention and separation selectivity are affected primarily by the following characteristics of the mobile phase: pH, choice of ion-pair agent, its concentration, and ionic strength. Some of these options are redun-

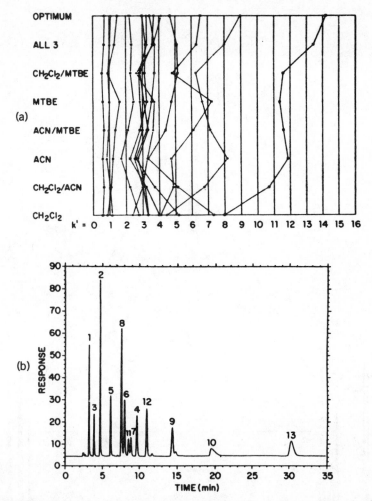

Figure 2.17   Change in Band Spacing for Normal-Phase Chromatography (Silica Column) as a Function of Strong Solvents in Mobile Phase. (a) k' vs. mobile phase for each compound; (b) separation of sample with optimum mobile phase. Reprinted with permission from Ref. (18).

dant since they overlap similar effects from other variables (20). For this reason, most of the band spacing control possible in ion-pair chromatography can be achieved by varying pH and the concentration of the ion-pair agent. This is illustrated in Fig. 2.18 for the case of a five-component mixture separated by ion-pair (IP) chromatography. The separation order is quite different for each of these three chromatograms: (a) pH 2.5, no IP agent; (b) pH

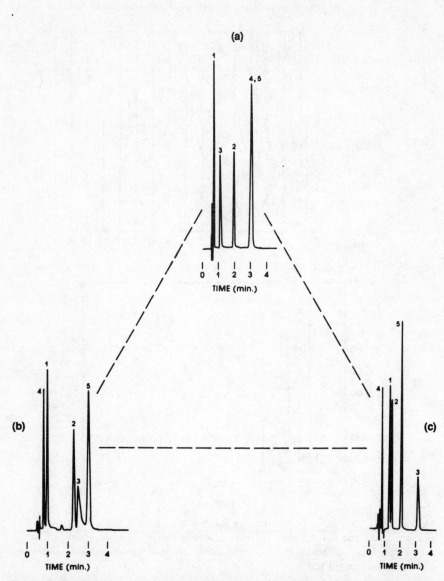

Figure 2.18 Effect of pH and Concentration of Ion-Pair Reagent on Band Spacing in Separation of a Five-Component Cold/Cough Preparation. Column, 15 × 0.46-cm, 5-μm C-8; 50° C; 3 mL/min; compounds are: 1, phenylephrine; 2, glycerol guaicolate; 3, pseudoephedrine; 4, sodium benzoate; 5, methylparaben. (a) mobile phase is 30% methanol/water, pH 2.5; (b) 27% methanol/water, pH 7.5; (c) 34% methanol/water, pH 5.5, 200 mM hexane sulfonate. Reprinted with permission from Ref. (20).

7.5, no IP agent; (c) pH 5.5, 200 mM IP agent. The optimum separation for this sample (not shown) provided sharp, well- separated bands for this sample (20).

Complexing Agents

Special selectivity effects can be created by the use of specific mobile-phase additives that complex reversibly with various sample compounds. One example is silver ion, which forms complexes with cis-olefins (21) and various compounds containing nitrogen (22). Other examples include borates for cis-diols (23) and certain crown ethers for primary amines (24). Complexing agents are rarely used in HPLC, because their mode of action is so specific. Usually, an appropriate complexing agent is not available that will change the band spacing for a particular sample. However, there are occasional applications where the use of chemical complexation results in a spectacular improvement in separation, as illustrated by the complex mixture of catecholamines in Fig. 2.19. In Fig. 2.19a, without the crown-ether additive, only five bands are resolved. In Fig. 2.19b, addition of the crown ether, 18-crown-6, selectively retards the elution of the six primary-amine derivatives, resulting in the resolution of all nine components. The poor resolution in Fig. 2.19a is in part due to a mobile phase that is too strong; some improvement in separation would have resulted from a simple decrease in the percent organic in the mobile phase.

2.4   RESOLUTION VS. CONDITIONS: COLUMN TYPE AND TEMPERATURE SELECTIVITY (α)

Column-Type Selectivity

A change in the nature of the stationary phase (column type) will generally affect both retention and α. Because stationary phase synthesis is often not well controlled, even columns of nominally the same type may show significant differences in selectivity from column to column. In this case, there is a problem in column-to-column reproducibility, as discussed further in Sect. 3.3.

A change in column selectivity is usually a composite of two effects. First, there is the intentional selectivity that results from the choice of a particular stationary phase: C-18, phenyl, cyano, silica, diol, etc. Second, there is an "accidental" selectivity that can result from the residual silanol groups that are always present in bonded-phase packings. Workers who intentionally vary the column to achieve better α-values are often making use of this second

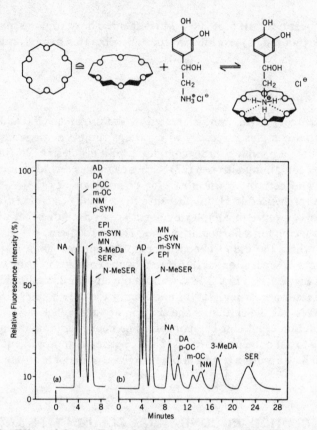

Figure 2.19   Effect of Crown-Ether Complexing Agent on Band Spacing of Primary and Secondary Amines by Reversed-Phase HPLC. NA, DA, p-OC, m-OC, NM, 3-MeDA, and SER are primary amines; (a) 0.01 M HCl mobile phase; (b) same, plus 5 g/L 18-crown-6. Reprinted with permission from Ref. (24).

kind of column selectivity. The origin of these silanol effects, and ways for limiting their contribution to column irreproducibility, are discussed in Chapter 3.

Bonded-Phase Selectivity

Replacing a C-18 column by a column with a different bonded-phase, such as phenyl or cyano, generally results in a change in band spacing. This is illustrated by the chromatograms in Fig. 2.20 for the separation of a five-component sample on three different columns. Conditions were otherwise the same

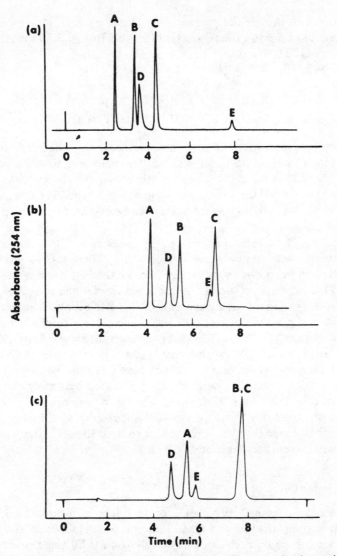

Figure 2.20 Column "Strength" and Band Spacing. Compounds: A, cortisone; B, dexamethasone; C, corticosterone; D, o-nitrophenol; E, fluorobenzene. (a) C-18 column, 36% acetonitrile/water; (b) phenyl column, 29% acetonitrile/water; (c) cyano column, 20% acetonitrile/water. Reprinted with permission from Ref. (25).

in each run, except that the percent acetonitrile/water in the mobile phase was adjusted to keep run times constant (8 min). In Fig. 2.20 the elution order changes as follows:

$$A < B < D < C < E \quad \text{(C-18)}$$
$$A < D < B < E < C \quad \text{(phenyl)}$$
$$D < A < E < B = C \quad \text{(cyano)}$$

These results suggest that different elution orders can be obtained simply by changing the column. This is not generally true, however, and this particular sample (Fig. 2.20) was specifically selected to show the ability of the stationary phase to affect α. Usually, the resulting changes in band spacing are more modest, and not nearly as spectacular as can be obtained by varying the mobile phase (Sect. 2.3).

An analysis of these column-selectivity effects as in Fig. 2.20 shows that they arise primarily from *column strength* (25). Thus, a C-18 phase is the least polar and functions as a stronger (more retaining) column in reversed-phase HPLC. A phenyl stationary phase is more polar and is weaker, while a cyano phase is still more polar and weaker. This effect is also seen in Fig. 2.20 as a result of the mobile phases used with each column to produce approximately equivalent k' ranges: C-18, 36% acetonitrile/water; phenyl, 29% acetonitrile/water; cyano, 20% acetonitrile/water. The selectivity of different reversed-phase columns operated in this fashion (constant run times, varying percent organic) is similar to the selectivity achieved by simply varying percent organic (and varying k'). However, when using column selectivity in this way, it is not necessary to worry as much about keeping the sample in the right retention range ($1 < k' < 20$), since column strength compensates for the effect of the mobile-phase strength on k'.

Silanol Selectivity

As discussed in Chapter 3, the silanols present in bonded-phase packings can vary both in concentration and type. Some silanols retain basic and cationic samples very strongly; such samples can show quite different retention and band spacing on columns of nominally the same kind but from different manufacturers. This effect is illustrated in Fig. 2.21 for the separation of a mixture of 16 cephalosporins. Here, there are numerous changes in separation order, as illustrated by bands 10, 12, and 13:

$$\text{(a, LiChrosorb C-8)} \quad 12 < 13 < 10$$
$$\text{(b, Zorbax C-8)} \quad 12 < 10 < 13$$
$$\text{(c, Partisil C-18)} \quad 10 < 13 < 12$$

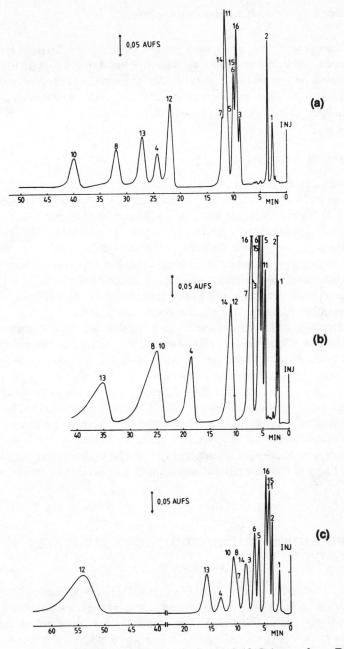

Figure 2.21 Column Selectivity Effects for C-8 and C-18 Columns from Different Manufacturers. Cephalosporin mixture. (a) LiChrosorb RP-8 column, 9% acetonitrile/buffer; (b) Zorbax C-8 column, 1% acetonitrile/buffer; (c) Partisil ODS column, 7% acetonitrile/buffer. Reprinted with permission from Ref. (26).

While changes in α among reversed-phase columns from different manufacturers appear to be an attractive way to control band spacing, this approach is accompanied by several disadvantages that are discussed further in Chapter 3. For this reason, *we do not recommend the use of different brands of C-8 or C-18 columns* (as in Fig. 2.21) *to vary band spacing*.

Temperature Selectivity

Temperature occasionally is used as a variable for controlling band spacing (27). However, when solvent strength is adjusted to compensate for decreased retention at higher temperatures (i.e., when run time is held constant), the effect on α is often small. Since the column plate number N usually is larger at higher temperatures, the maximum resolution will usually occur at higher temperatures regardless of band spacing (28). For this reason, lowering the temperature seems less useful for changing values of α.

There are two general cases where a change in temperature can be expected to affect band spacing significantly. First, theory suggests that compounds having a different molecular shape or size will show changes in α as temperature is varied (29,30). However, most samples to be separated by HPLC are structurally related, so that their shapes are similar. Consequently, shape-related temperature selectivity generally is not useful as a variable for controlling band spacing. The second case, where a change in temperature can affect band spacing, involves samples that are partly ionized under the conditions of separation. Because the degree of ionization is temperature dependent and different for different compounds, the usual pattern of constant band spacing with change in temperature may be broken. This effect is particularly likely in ion-pair separations and will be considered further in Chapter 5.

2.5   RESOLUTION VS. CONDITIONS: COLUMN PLATE NUMBER (N)

The column plate number plays an important role in determining sample resolution (Eqn. 2.3a). Usually, method development begins with a high-performance column that can exhibit 10,000–20,000 plates under appropriate conditions, e.g., a 10–25 cm column with 3–5 μm particles. We recommend a 25-cm, 5-μm column. Once solvent strength and band spacing have been optimized (Sects. 2.2–2.4), conditions can be adjusted to provide whatever value of N is required for the required resolution. However, if values of α are quite small, this may mean that a very large N-value and a corresponding long run time will be necessary. The following rough guidelines show required values of N and run time as a function of α, for a minimum resolution of $R_s = 1.5$:

α	Column	N^a	Run Time
1.10	4 cm, 3 μm	6,000	2–5 min
1.05	30 cm, 5 μm	25,000	30–60 min
1.02	5 m, 10 μm	160,000	8–15 hr

[a]Calculated as described in Equation 2.3.

Practical separations (run times less than 1 hr) require α-values of at least 1.05, suggesting that N should not be optimized until the minimum α-value exceeds 1.05.

Column Conditions

The theory of how values of N vary with experimental conditions is now well developed (1,3,31). Once the column packing, temperature, and mobile phase have been selected, N can be controlled by selecting column length and particle size and/or changing the flow rate. We will refer to these latter variables as *column conditions*. Other factors being equal, N will increase in proportion to column length and with decreasing particle diameter. For each set of experimental conditions, there is an optimum flow rate—typically about 1 mL/min for particles of 3–5 μm (assumes 0.4–0.5 cm i.d. column). Higher flow rates lead to lower values of N, as illustrated in Fig. 2.22 for columns with different dimensions. Values of N also decrease as the mobile-phase viscosity is increased; therefore, low-viscosity solvents should be used where feasible. For reversed-phase systems, the higher viscosity of water means that there will usually be an advantage in operating at higher temperatures, to maximize values of N. A temperature of 50° C represents a good compromise in this connection.

When adjusting the column plate number to achieve the desired resolution of the sample, attention must be paid to certain other method performance factors:

- Run time   Should be as short as possible, preferably less than 10–20 min
- Pressure   Pressures less than 150 bar are desirable
- Peak heights   Taller, narrower peaks may be required whenever detection sensitivity is limited (Sect. 4.1)
- Solvent usage For routine, high-volume tests, a reduction in the mobile phase consumed per run is desirable, but less important

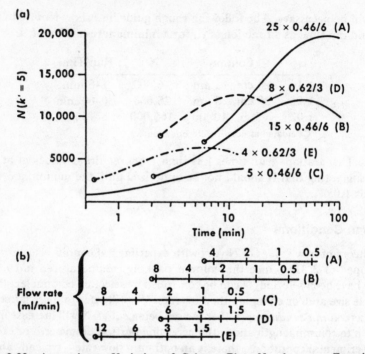

Figure 2.22 Approximate Variation of Column Plate Number as a Function of Column Conditions. (a) Plate number vs. time for columns of different geometry (e.g., curve (A) is for a 25 × 0.46 cm column of 6 μm particles.) (b) Flow rates corresponding to plots of (a). Reprinted with permission from Ref. (32).

Optimizing parameters such as N, resolution, run time, pressure, etc., for various column conditions can be approached by trial and error. Rough guidelines for estimating suitable column conditions to achieve a required N-value or sample resolution are given in Ref. 32, along with advice on minimizing run time and pressure. Figure 2.22 summarizes these recommendations.

While the same plate number N can be achieved by different combinations of column conditions, some conditions produce faster separations. Thus, when the required N-value is small (α is large), shorter columns of 3-μm particles are preferred. When the required value of N is large, a 25-cm, 5-μm column is preferable. Columns of 3-μm particles usually provide a larger number of plates per unit time than do columns with larger particles. This situation is illustrated in Fig. 2.23 for the reversed-phase separation of a mixture of phenols; a 3-μm column (Fig. 2.23b) provides slightly better separation in about half the time required for a 6-μm column (Fig 2.23a). On the other hand, 3-μm columns are somewhat more difficult to work with and become plugged more easily.

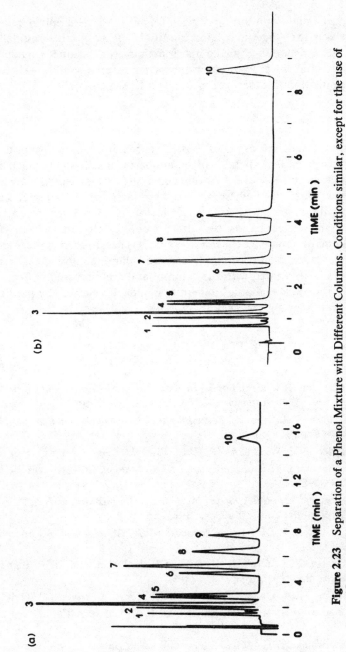

Figure 2.23 Separation of a Phenol Mixture with Different Columns. Conditions similar, except for the use of a 15-cm, 6-μm column in (a) and an 8-cm, 3-μm column in (b). Reprinted with permission from Ref. (33).

The required plate number for a given separation is generally obtainable through systematic changes in column conditions, using the above guidelines. However, after the mobile-phase composition has been selected, a much simpler and more efficient procedure for optimizing column conditions is to use computer simulation as described in Refs. (3,4); see also Sect. 8.4.

Nonideal Separation

The preceding discussion assumes that the chromatographic system is well behaved, in which case predictable plate numbers will be found for each band in the chromatogram. However, it is often observed that measured plate numbers are lower than expected, sometimes by a factor of 10 or more. At the same time, tailing bands with asymmetry factors $A_s > 1.5$ may be observed ($A_s = 1$ for symmetrical bands; see Glossary of Symbols and Terms). When these symptoms of nonideal behavior are seen, it is best to correct the underlying problem, rather than continue to explore column conditions. These effects are usually associated with the column, and are discussed in Sect. 3.3. Computer software is available to identify nonideal behavior, by predicting the ideal N-value for a given peak (Sect. 8.4).

REFERENCES

1. L. R. Snyder and J. J. Kirkland, *An Introduction to Modern Liquid Chromatography*, 2nd ed., Wiley-Interscience, New York, 1979, Chapt. 2.
2. C. F. Poole and S. A. Schuette, *Contemporary Practice of Chromatography*, Elsevier, Amsterdam, 1984, p. 9.
3. L. R. Snyder and J. W. Dolan, *Am. Lab. (Fairfield, Conn.)*, August 1986, p. 37.
4. L. R. Snyder and J. W. Dolan, *DryLab I User's Manual*, LC Resources Inc., San Jose, Calif., 1987.
5. J. C. Berridge, *Techniques for the Automated Optimization of HPLC Separations*, Wiley-Interscience, New York, 1985, pp. 21–27.
6. M. A. Quarry, E. I. Du Pont de Nemours, Co., Wilmington, Del., unreported data (1986).
7. M. A. Quarry, R. L. Grob, L. R. Snyder, J. W. Dolan, and M. P. Rigney, *J. Chromatogr.*, *384* (1987) 163.
8. L. R. Snyder, M. A. Quarry, and J. L. Glajch, *Chromatographia, 24* (1987) 33.
9. P. J. Schoenmakers, H. A. H. Billiet, and L. de Galan, *J. Chromatogr.*, *185* (1979) 179.
10. P. J. Schoenmakers, H. A. H. Billiet, and L. de Galan, *J. Chromatogr.*, *218* (1981) 259.

11. W. R. Melander and Cs. Horvath, in *High-Performance Liquid Chromatography. Advances and Perspectives, Vol 2*, Cs. Horvath, ed., Academic Press, New York, 1980, p. 114.

12. L. R. Snyder, *J. Chromatogr.*, *92* (1974) 223.

13. L. R. Snyder, *J. Chromatogr. Sci.*, *16* (1978) 223.

14. J. L. Glajch, J. J. Kirkland, K. M. Squire, and J. M. Minor, *J. Chromatogr.*, *199* (1980) 57.

15. L. R. Snyder, in *High-Performance Liquid Chromatography. Advances and Perspectives, Vol 3*, Cs. Horvath, ed., Academic Press, New York, 1983, p. 157.

16. L. R. Snyder, J. L. Glajch, and J. J. Kirkland, *J. Chromatogr.*, *218* (1981) 299.

17. M. de Smet, G. Hoogewijs, M. Puttemans, and D. L. Massart, *Anal. Chem.*, *56* (1984) 2662.

18. J. L. Glajch, J. J. Kirkland, and L. R. Snyder, *J. Chromatogr.*, *238* (1982) 269.

19. M. T. W. Hearn, ed., *Ion-Pair Chromatography*, Marcel Dekker, New York, 1985.

20. A. P. Goldberg, E. Nowakowska, P. E. Antle, and L. R. Snyder, *J. Chromatogr.*, *316* (1984) 241.

21. R. J. Tscherne and G. Capitano, *J. Chromatogr.*, *136* (1977) 337.

22. L. A. d'Avila, H. Colin, and G. Guiochon, *Anal. Chem.*, *55* (1983) 1019.

23. L. J. Morris, *J. Chromatogr.*, *12* (1963) 321.

24. M. Weichmann, *J. Chromatogr.*, *235* (1982) 129.

25. P. E. Antle and L. R. Snyder, *LC Mag.*, *2* (1984) 840.

26. I. Wouters, S. Hendrickx, E. Roets, J. Hoogmartens, and H. Vanderhaeghe, *J. Chromatogr.*, *291* (1984) 59.

27. D. E. Henderson and D. J. O'Connor, *Advances in Chromatography*, *23* (1984) Chapt. 2.

28. J. R. Gant, J. W. Dolan, and L. R. Snyder, *J. Chromatogr.*, *185* (1979) 153.

29. L. R. Snyder, *J. Chromatogr.*, *179* (1979) 167.

30. J. Chmielowiec and H. Sawatzky, *J. Chromatogr. Sci.*, *17* (1979) 245.

31. R. W. Stout, J. J. DeStefano, and L. R. Snyder, *J. Chromatogr.*, *282* (1983) 263.

32. L. R. Snyder and P. E. Antle, *LC Mag.*, *3* (1985) 98.

33. R. W. Stout, J. J. DeStefano and L. R. Snyder, *J. Chromatogr.*, *261* (1983) 189.

3

THE ROLE OF
THE COLUMN

The importance of the column in HPLC separation needs no elaboration. While high-performance columns are often taken for granted, commercial columns can differ widely among different suppliers (e.g., Fig. 2.21), and even between nominally identical columns from a single source. These differences can have a serious impact on method development in HPLC, so we will examine this area in some detail. Specifically, different columns can vary in plate number, band shape (tailing), retention and selectivity, and lifetime.

The importance of the column plate number N for controlling resolution (Eqn. 2.3) was discussed in Chapter 2. However, the large N-values that are readily attained with, for example, a 15-cm, 5-μm column, may be taken for granted. Many workers ignore the possibilities of varying the following column conditions: column dimensions, particle size, and flow rate. These experimental variables do not affect relative retention (k′ and α values), but they do affect N. In many cases, it is possible to further improve a separation, after optimizing selectivity, by altering column conditions (see Sect. 2.5). In this way, resolution can be improved, separation time decreased, column pressure minimized, detectability improved (by sharpening bands), and solvent consumption reduced.

This chapter will review what is known about different columns and column packings. We will discuss possible problems in the use of the column, ways these problems impact method development, and various remedies. We also will consider the role of "good" columns in optimizing a routine HPLC procedure.

3.1 CHARACTERISTICS OF COLUMNS AND COLUMN PACKINGS

Packing Particles

HPLC column packings are either polymer- or silica-based. Because silica-based packings are used for most separations, their use is emphasized in this book. Unless otherwise noted, we will discuss particles formed from a silica matrix, bonded with an organic surface layer such as C-8 or C-18 (see Ref. 1). While the chemical nature of these packings varies with the stationary phase (e.g., C-8), their physical characteristics are similar.

The average particle diameter of HPLC packings typically is between 3 and 10 μm, with a narrow size distribution. Both spherical and irregular shapes are available, and in most cases, the particles are totally porous with a relatively uniform pore structure. The reproducibility and long-term stability of the HPLC column are of critical importance in method development, particularly for analytical methods that will be used over a long time and in different laboratories.

Irregularly-shaped silica particles in the 5–10 μm range are produced by grinding and sizing bulk silica. Today, these particles are not as popular, because of difficulties in preparing and maintaining efficient columns. Also, column packings based on irregular materials tend to be less mechanically stable than spherical particles, and column back pressures may be higher because of fragmentation of these particles to produce "fines."

Most HPLC packings are based on spherical silica particles having good mechanical strength and a narrow particle-size distribution. Methods for preparing such particles are summarized in Ref. (2). Figure 3.1 illustrates the formation of Zorbax® silica microspheres (DuPont) by aggregating nonporous, colloidal silica into porous particles of uniform size (3).

Silica particles from different manufacturers are prepared by various procedures. These materials can exhibit different chromatographic properties

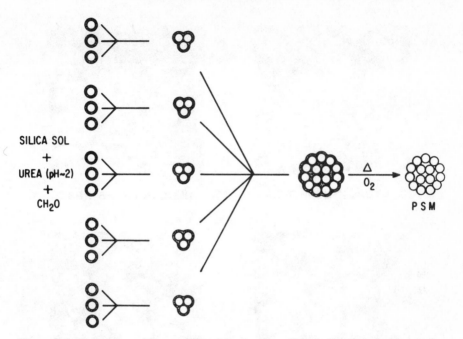

Figure 3.1 Formation of Porous Silica Microspheres (PSM). Reprinted with permission of DuPont Medical Products Department.

due to variations in purity, surface area, pore-size distribution, and surface chemistry. Figure 3.2 compares the visual appearance (shape and particle-size distribution) of some commercially available silica particles.

For best results, the particle-size distribution should vary by no more than $\pm50\%$ from the mean particle size. As discussed later (also see Figs. 4.7 and 4.10), columns of smaller particles (e.g., 3 μm) permit faster separations compared with columns of larger particles (e.g., 6 μm). However, 5-6 μm particles generally represent a good compromise in terms of convenience, performance, and column lifetime.

The strength of the silica particles affects how the HPLC column should be packed. This, in turn, determines the ultimate performance and stability of the column. Higher-strength particles provide columns that exhibit lower back pressures and longer lifetimes (5).

The choice of pore size and surface area of the column packing is dictated by the application. Small molecules are usually chromatographed on particles that have 7-12 nm pores, while larger molecules generally are separated with packings having larger pores (e.g., 30 nm for proteins). For good results, the pore diameter should be at least three times larger than the diameter of

VYDAC™-214-TP5

ULTRAPORE™-5-RPSC

ZORBAX®-PSM-300

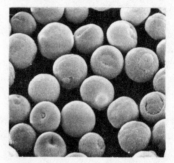

NUCLEOSIL™-7-300

⊢⎯⎯⎯⎯⎯⎯⎯⊣
10µm

Figure 3.2   Transmission Electron Micrographs of Some Porous Silica Microparticles. Reprinted with permission from Ref. (4).

the sample molecules (6). This configuration permits access of sample molecules to the interior of the column-packing particle, and also allows relatively rapid diffusion of sample molecules within the pore.

The chemical nature of the silica surface is particularly important, and is affected by differences in manufacturing procedures. The silica surface contains various kinds of silanol (SiOH) groups, as illustrated in Fig. 3.3. Silica heated above 800° C is largely devoid of silanol groups (Fig. 3.3a); such a material is of little value in HPLC. As shown in Fig. 3.3b, the surface of the silica can be fully hydroxylated (maximum silanol concentration of 8.0 μmol/m^2). Silica packings sometimes exist in a partially hydroxylated state (5–7 μmol/m^2 silanol) (7).

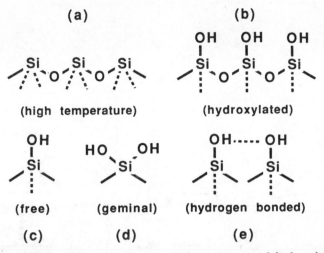

Figure 3.3   Silica Surface Chemistry. (a) High-temperature dehydroxylated silica; (b) fully-hydroxylated (ideal) silica; (c–e) different silanol types.

Individual silanols exist as three general types, as suggested in Figs. 3.3c–e (see also Ref. 8). The silanol groups on a fully-hydroxylated silica occur mainly in a hydrogen-bonded or geminal form (Figs. 3.3d,e) (7). Free or non-hydrogen-bonded SiOH groups (Fig. 3.3c) occur in relatively low concentrations; however, these silanols can cause strong binding of basic solutes due to their highly acidic nature. Free silanols are, therefore, undesirable for the separation of basic samples. Partially-hydroxylated silicas apparently contain a higher relative concentration of these free, more acidic SiOH groups.

Packings that contain more acidic silanols exhibit increased retention and broad, tailing bands for *basic* samples. Other packings can show increased retention and broad, tailing peaks for *acidic* samples. Several examples of band tailing due to these silanol effects are illustrated in Fig. 3.4. The same mixture of three acidic components, two basic compounds, and a neutral compound was injected for each run. Two reversed-phase columns were used, LC-18 and LC-18-DB; the latter column has been processed for reduced retention and peak tailing of basic compounds. Mobile-phase conditions were also varied.

Figures 3.4a and b show the separation on each column using a mobile phase of 7%v acetonitrile/phosphate buffer, pH 3.5. In Fig. 3.4a, the basic compounds procainamide (PA) and N-acetylprocainamide (NAPA) are strongly retained and tail so badly that they are barely observable above the baseline on the LC-18 column. Similar behavior is common for basic compounds on many reversed-phase HPLC columns. In Fig. 3.4b, these basic

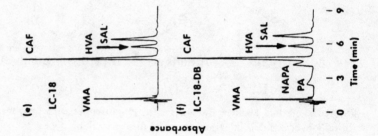

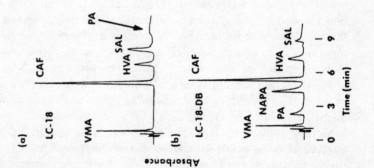

58

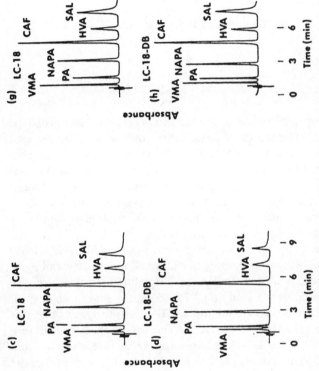

Figure 3.4 Secondary Retention Effects in the Reversed-Phase Separation of Neutral, Basic, and Acidic Compounds. Columns: 15 × 0.46-cm Supelcosil LC-18 (a,c,e,g) and LC-18-DB (b,d,f,h); mobile phase: 7%(v) acetonitrile/phosphate buffer (pH 3.5), 10 mM TEA added in (c,d,g,h) and 1% acetate added in (e–h). Peaks: (neutral) caffeine (CAF); (acidic) homovanillic acid (HVA), vanillylmandelic acid (VMA), and salicylic acid (SAL); (bases) procainamide (PA) and N-acetylprocainamide (NAPA). Reprinted with permission from Supelco, Inc.

compounds elute with decreased retention and improved peak shape on the LC-18-DB column; however, the separation is still not satisfactory. The increased retention and tailing of these basic compounds in Fig. 3.4a suggests that the silanols of the LC-18 column are more acidic (or more accessible) than those on the LC-18-DB column (Fig. 3.4b).

The strong retention and peak tailing of basic compounds such as PA and NAPA in reversed-phased systems are apparently the result of two kinds of silanol interaction:

hydrogen bonding

$$R_3N: \ + \ HO\text{-}Si \ \rightleftharpoons \ R_3N: \cdots H \cdots O\text{-}Si\text{-}$$

ion exchange

$$R_3NH^+ \ + \ (X^+)\text{-}O\text{-}Si \ \rightleftharpoons \ (R_3NH^+)\text{-}O\text{-}Si\text{-} \ + \ X^+$$

The addition of appropriate mobile-phase modifiers can eliminate problems due to these secondary interactions of basic compounds with acidic silanols. Various tertiary amines such as triethylamine (TEA) and morpholine in concentrations of 10–50 mM are effective for this purpose. These modifiers preferentially block the silanols that cause secondary retention. Figures 3.4c and d show the results of adding 10 mM TEA to the mobile phase for the LC-18 and LC-18-DB columns, respectively; the bands for the basic compounds PA and NAPA are now much sharper, with little tailing.

In Figs. 3.4a–d, peaks for the acidic compounds homovanillic acid (HVA) and salicylic acid (SAL) vary considerably in width and asymmetry; tailing of SAL is particularly pronounced on the LC-18-DB column. In this case, the tailing of SAL is not improved by adding TEA (compare Figs. 3.4b and d), which blocks the effect of acidic silanol groups. This result suggests that different kinds of silanols may be present on the surfaces of these two column packings; the silanols on the LC-18 column appear to preferentially bind bases, while the LC-18-DB silanols seem to interact strongly with carboxylic acids. The interaction of acids with the silica surface is greatly reduced by deliberately adding a small amount of a carboxylic acid to the mobile phase. This approach is illustrated in Figs. 3.4e (LC-18) and f (LC-18-DB), where 1% acetate was added to the original mobile phase used in Figs. 3.4a and b (pH constant, no base added). The acidic compounds HVA and SAL now elute as sharper bands with no tailing. While the addition of acetate to the mobile phase (separations of Figs. 3.4e and f) improves peak shape for the acidic compounds, the basic compounds show slightly poorer peak shape and increased retention as a result of acetate addition.

For samples with both acidic and basic compounds, both acidic and basic modifiers can be added to the mobile phase. In Figs. 3.4g and h, 1% acetate and 10 mM TEA were added to the original mobile phase. Now, all bands in the chromatograms for both columns (LC-18 and LC-18-DB) are symmetrical, and the retention of each solute is almost identical for both columns. The suppression of secondary retention effects in this case appears to have eliminated all retention differences (column-to-column variability). Note that peak shape and retention for the neutral solute caffeine (CAF) remains essentially unchanged in Figs. 3.4a–h, suggesting an absence of silanol effects for this compound. For a more detailed discussion of silanol-related problems and their correction, see Ref. (9).

A major factor in the lack of uniformity between column packings (as in Fig. 3.4) is differences between the silica supports. Selectivity changes observed with nominally the same bonded-phase column from various manufacturers (e.g., C-18) are largely a function of variations in the silica substrate, rather than differences in the organic bonded phase. Depending on the type of silica, reversed-phase columns can be classified as basic or acidic. For example, the LC-18 column of Fig. 3.4 can be considered to be acidic, while the LC-18-DB column is relatively basic.

Silicas with good properties towards basic compounds (e.g., Supelcosil LC-18-DB) have apparently been fully hydroxylated, so as to provide a maximum concentration of homogeneous hydrogen-bonded SiOH groups (5). Table 3.1 ranks some commercial packings according to their "goodness" for separating basic and acidic samples. Packings at the top of this list are less acidic and are preferred for separating basic compounds; packings at the bottom of the list are more acidic, and often are preferred for separating acidic components.

Bonded Phases

Most bonded-phase, silica-based HPLC packings are made with surface-reacted organosilanes, using the reactions shown in Fig. 3.5. The most commonly used process (Fig. 3.5a) involves the reaction of chlorodimethylsilanes with surface silanol groups. Various alkyl and substituted alkyl silicas for reversed-phase HPLC are made by this reaction. The bonded-phase packings that result from monofunctional reactions as in Fig. 3.5a are definable and reproducible, and packings made by this route often exhibit the largest N-values.

A few packings use a polymerized surface layer that results from the reaction of di- or trifunctional silanes with the silica particle (Figs. 3.5b and c). The polymeric bonded-phases obtained from such reactions appear to be somewhat more stable; however, this reaction can be less reproducible, and

TABLE 3.1   Ranking of Commercial Columns According to Increasing Acidity[a]

Relative "Goodness" for Basic Compounds	
Zorbax Rx	(good)
Vydac	
Rsil	
Nucleosil	
Polygosil	
Novapak	
μ-Bondapak	
Supelcosil DB	
Spherisorb 2	
LiChrosorb	
Chrompack	
Hypersil	
Perkin-Elmer	
Supelcosil	
Zorbax	
MicroPak	(poor)

Source: Organized mainly from Refs. 10–12.
[a]Differences between successive listings may not be significant.

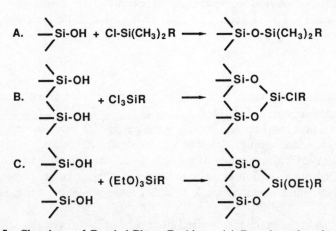

Figure 3.5   Chemistry of Bonded-Phase Packings. (a) Reaction of surface silanol with chlorodimethylsilane, (b) Reaction of surface silanols with trifunctional chlorosilane, (c) Reaction of surface silanols with trifunctional alkoxysilane.

the resulting packings may be more variable (compared with monofunctional phases, Fig. 3.5a) with respect to retention and selectivity.

The stability of bonded-phase packings is especially important in method development. Once the desired separation method has been obtained, column characteristics should remain unchanged for as long as possible, so that any need for further adjustment of separation conditions (or replacement of the column) is minimized. When used under the same conditions, long-chain alkyl-bonded-phase packings (e.g., C-18 and C-8) generally are more stable than short-chain bonded phases (e.g., C-3 and C-4); polymeric phases tend to be more stable than monomeric phases. Nonpolar bonded phases (e.g., C-4) usually are more stable than polar phases (e.g., diol). Some studies suggest that substitution of sterically-protecting substituents (e.g., isopropyl) for the dimethyl groups on the silane silicon (Fig. 3.5a) can significantly improve the stability of reversed-phase packings against degradation by aggressive mobile phases (13). Columns based on this technology are now commercially available as Du Pont's Zorbax® Rx Series.

End-capping is often used to more completely bond (silanize) the packings shown in Figs. 3.5a and b. End-capping consists of a subsequent reaction with trimethylchlorosilane or (less often) hexamethyldisilazane, to increase the coverage of the support, and to minimize unwanted interaction with silanols (as in Fig. 3.4). However, end-capping cannot completely overcome the disadvantages of an acidic silica (Table 3.1). In addition, the TMS group is readily hydrolyzed from the packing in reversed-phase separations, making this approach of marginal merit for many long-term applications (4).

A third commonly used reaction for bonded-phase packings involves the reaction of alkoxy- (or siloxy-) silanes with silanol groups on the silica surface (Fig. 3.5c). This alkoxy reaction is typically carried out with R groups (Fig. 3.5) that would react with chlorosilanes (e.g., amino or diol phases). The alkoxy reaction can be performed with mono-, bi-, or trifunctional silanes, with the same advantages and limitations described above for the chlorosilanes.

Variations in retention and selectivity due to differences in the bonded phase (apart from differences in the silica support) can arise from several sources:

(a) The choice of silane—monofunctional or polyfunctional
(b) The completeness of bonding—partially or fully reacted
(c) The presence or absence of end-capping
(d) The bonding chemistry itself

Each of these factors can have a significant effect on the final results from method development. Separations with columns made from monofunctional silanes often are more reproducible from batch to batch, when compared with columns prepared from polyfunctional silanes. Fully-reacted packings (high bonded-phase concentrations) also are more reproducible (and sometimes more stable), when compared with packings with partially-reacted (lower silane-concentration) surfaces. Significant selectivity differences can be seen for fully vs. partially-reacted packings of the same bonded phase (e.g., C-18). Selectivity and peak shape differences also can occur between packings that have been end-capped compared with those that have not, depending on the particular sample being separated. Finally, differences in the silanization reaction conditions can result in retention variations between bonded-phase packings of nominally the same phase. Most workers prefer monofunctional bonded-phase packings that are fully reacted and end-capped.

Reversed-phase separations also can be carried out with columns made of spherical, porous particles made (usually) from polystyrene. Such "polymeric" (nonsilica) particles are stable from pH 1 to 13 allowing their use at higher pH-values (>8), where silica-based columns degrade rapidly. There are some reports that polymeric reversed-phase columns give better results for certain basic compounds, compared to silica-based columns. For example, Fig. 3.6 shows the separation of some basic antibiotics using both silica-based and polymeric packings. While polymeric particles appear to surmount certain limitations of silica-based packings, there is a much greater choice of different silica-based packings (column and particle size, different bonded phases, etc.). Silica-based columns also are (usually) somewhat more efficient than polymeric columns. These factors, plus the greater experience with silica-based packings, usually make the latter materials more attractive as a starting place for method development.

Some useful column-packing types for the different HPLC methods are listed in Table 3.2.

3.2   COLUMN SPECIFICATIONS

The specific requirements for a given separation usually determine the type and configuration of the column to be used (particle size, length, internal diameter, etc.). Section 2.5 discusses the value of having columns with different lengths and particle sizes. There are normally many possible suppliers for a given type of column; however, these columns can vary greatly in performance. Therefore, certain information concerning column specifications and performance is required for use in method development and subsequent routine operation. Column specifications of interest include:

(a) Minimum plate number N for a given value of k'

(b) Maximum peak asymmetry factor A_s

(c) Retention (k') reproducibility

(d) Typical pressure drop

(e) Bonded-phase concentration (if applicable)

These data can be obtained from the column manufacturer prior to purchase. In addition, some manufacturers provide actual data of this type for each column sold.

Before discussing column specifications, we need to consider the general nature of HPLC columns used during method development. Columns are usually 5–25 cm in length, packed with 3–10 μm particles. The internal diameter (i.d.) of most analytical columns is 0.4–0.5 cm. "Microbore" columns (0.1–0.2 cm i.d.) are available for special applications involving samples available in limited quantity (e.g., trace analysis, see Sect. 7.1), or for interfacing with detectors such as a mass spectrometer (15). Wider columns are used for size-exclusion chromatography, and for very small-particle columns to minimize extra-column band broadening (16). Column blanks are generally made of straight, stainless-steel tubes terminated at the ends with compression fittings. Short (5–10 cm) "cartridge" columns of metal, glass, or plastic have recently become popular, since they are designed with easy-connect end-fittings that permit (a) convenient replacement and (b) the connecting of individual units into longer lengths when required. These cartridge columns are becoming more popular, because they are less costly and are easily connected in series to provide an optimum column length for a particular separation (see Sects. 2.5 and 9.4 for selecting column conditions).

Plate Number

The column plate number N is the single most important characteristic of a column. N defines the ability of the column to produce sharp, narrow peaks and to achieve good resolution for band-pairs with small α-values. The measurement of column plate number is discussed in Sect. 2.5. Table 3.3 shows typical plate numbers (small molecules, MW ~ 200) for well-packed HPLC columns of various lengths and particle sizes (see also Fig. 2.23). Note that these values are obtained under "optimum" conditions—small, neutral solutes, low viscosity mobile phases, and a low flow rate. Most manufacturers specify the conditions they use to measure N; this test should be repeated when the column is first received. If the plate number is significantly lower

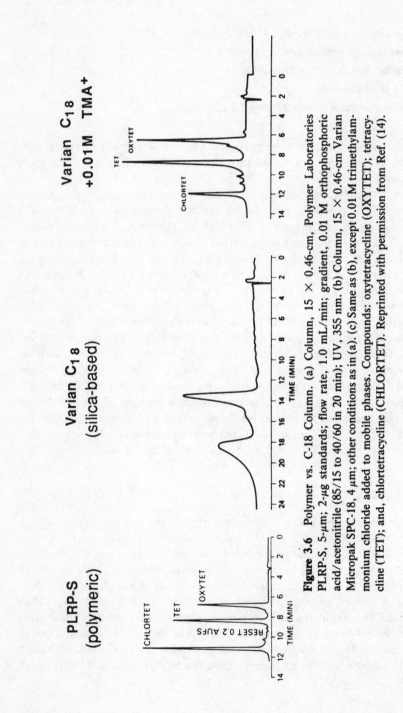

Figure 3.6 Polymer vs. C-18 Column. (a) Column, 15 × 0.46-cm, Polymer Laboratories PLRP-S, 5-μm; 2-μg standards; flow rate, 1.0 mL/min; gradient, 0.01 M orthophosphoric acid/acetonitrile (85/15 to 40/60 in 20 min); UV, 355 nm. (b) Column, 15 × 0.46-cm Varian Micropak SPC-18, 4 μm; other conditions as in (a). (c) Same as (b), except 0.01 M trimethylammonium chloride added to mobile phases. Compounds: oxytetracycline (OXYTET); tetracycline (TET); and, chlortetracycline (CHLORTET). Reprinted with permission from Ref. (14).

TABLE 3.2 Useful Column Packings for HPLC[a]

Comments	Description
REVERSED-PHASE (AND ION-PAIR) METHOD	
C-18 (octadecyl or ODS)	Rugged; highly retentive; widely available
C-8 (octyl-)	Similar to, but slightly less retentive, than C-18
C-3, C-4	Less retentive; used mostly for peptides and proteins
C-1 (trimethyl-silyl, TMS)	Least retentive; least stable
phenyl, phenethyl	Moderately retentive; some selectivity differences
CN- (cyano)	Moderately retentive; used for both reversed- and normal-phase
NH_2- (amino)	Weak retention; used for carbohydrates; less stable
Polystyrene[b]	Stable with $1 < pH < 13$ mobile phases; better peak shape and longer column life for some separations
NORMAL-PHASE METHOD	
CN- (cyano)	Rugged; fairly polar; general utility
OH- (diol)	More polar than CN-
NH_2- (amino)	Highly polar; less stable
Silica[b]	Very rugged; cheap; less convenient to operate; used in prep LC
SIZE-EXCLUSION METHOD	
Silica[b]	Very rugged; adsorptive
Silanized silica	Less adsorptive, wide solvent compatibility; used with organic solvents
OH- (diol)	Less stable; used in aqueous SEC (gel filtration)
Polystyrene[b]	Used widely for organic SEC (GPC); incompatible with highly polar solvents
ION-EXCHANGE METHOD	
Bonded-phase	Less stable and reproducible
Polystyrene[b]	Less efficient; stable; more reproducible

[a]Silica-based bonded phases, except as noted.

[b]No bonded phase on these packings.

TABLE 3.3 Plate Number for Well-Packed, Small-Particle HPLC Columns

Particle Diameter (μm)	Column Length (cm)	Plate Number N
10	15	6,000–7,000
10	25	8,000–10,000
5	10	7,000–9,000
5	15	10,000–12,000
5	25	16,000–20,000
3	5	6,000–7,000
3	8	10,000–11,000
3	10	12,000–14,000

(N < 80% of the claimed value), the column should be returned to the manufacturer for replacement or refund.

For the case of an already developed HPLC method, the plate number of a new column should be determined for the particular sample of interest, under the conditions to be used for a specified separation (HPLC method). Since the column plate number is dependent on specific experimental factors, its value for compounds of interest may be smaller than the optimum value measured for a standard compound such as toluene. For large solutes in relatively viscous mobile phases, the value of N may be only a fraction (e.g., one-third) of the optimum value. Secondary retention for some solutes (i.e., from silanol effects) also can cause broader peaks and a smaller-than-optimum plate number. Such effects should be identified and resolved before starting method development (see Sect. 2.5). The addition of certain mobile-phase modifiers can often correct this problem, as discussed in Sect. 3.3.

A systematic record of N-values vs. time should be maintained for compounds of interest, so that column efficiency is known at any time. This record will help the operator to monitor column performance and to anticipate when column replacement (or repair) will be necessary. A record of N-values vs. time is also useful for judging the overall performance of columns from a particular manufacturer, and for other purposes.

Many practitioners prefer 15- or 25-cm columns of 5-μm particles as a starting point for method development. This configuration provides a large enough N-value for most separations, and such columns are today quite reliable. If a larger N-value is required for a particular separation, additional column lengths can be connected with low-volume fittings. Short columns of 3-μm particles are useful for carrying out very fast separations (e.g., <5 min.). However, columns of <5-μm particles generally are less suited for routine applications, since they are (a) more susceptible to sampling problems,

(b) more operator dependent, and (c) more affected by instrumental band-broadening effects.

Peak Asymmetry

While the column plate number is a useful measure of column quality, the shape of the peaks produced by the column can be equally important in method development. Columns and experimental conditions that provide symmetrical, Gaussian peaks are always preferred. Peaks with poor symmetry can result in (a) imprecise quantitation, (b) degraded resolution and undetected minor bands in the peak tail, and (c) problems with retention reproducibility. The degree of peak tailing or asymmetry is measured by the peak asymmetry factor A_s, calculated as in Figure 3.7. "Good" columns produce peaks with A_s values of 0.9–1.1 (exactly symmetrical peaks have an asymmetry factor of 1.0). Figure 3.8 illustrates the effect on shape and width of bands for a range of peak-asymmetry-factor values. Manufacturers often specify A_s values of 0.9–1.2 for new columns. Samples of interest should have A_s values of < 1.3 for a particular column being considered for method development (you may have to accept $A_s = 1.5$ in unusual cases, but no larger!) Figures 3.4g and h illustrate peaks with A_s values of about 1.0 for a reversed-phase

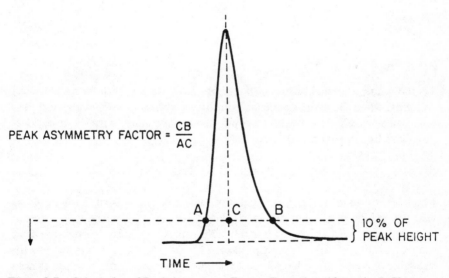

Figure 3.7   Calculation of Peak Asymmetry Factor. Reproduced from the *J. Chromatogr. Sci.* (Ref.17) by permission of Preston Publications.

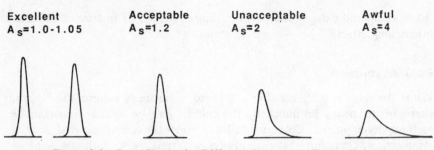

Excellent	Acceptable	Unacceptable	Awful
$A_s=1.0-1.05$	$A_s=1.2$	$A_s=2$	$A_s=4$

Figure 3.8 Peak Shapes for Different Asymmetry Factor Values.

system containing mobile-phase additives to improve both peak symmetry and retention reproducibility.

Retention Reproducibility

The reproducibility of retention times or k'-values among different columns can be further specified by chromatographing a series of standards, preferably including both polar and nonpolar molecules. Similar retention times (or k'-values) should be found for the test compounds when run under standard conditions. Few manufacturers supply information of this type, so in most cases such testing will have to be done by the user.

Pressure Drop

Similar column permeabilities or back pressures are expected for well-packed columns, when the same operating conditions, column dimensions, and particle size are used. The pressure drop for columns packed with spherical particles can be approximated by:

$$P = 3500 \, L\eta/t_o d_p^2 \qquad (3.1)$$

where, P is pressure in psi, L is column length in cm, η is the mobile-phase viscosity in cPoise, t_o is column dead-time in sec, and d_p is particle diameter in micrometers. New, spherical-particle columns should have a pressure drop that is within ± 30–50% of that predicted by Eqn. 3.1. Columns packed with irregular particles may give somewhat higher pressures. Some suppliers normally report a back pressure that was measured for a particular column under certain operating conditions.

Bonded-Phase Concentration (Coverage)

Well-made bonded-phase columns have a dense population of organic groups attached to the surface of the silica support. The actual coverage depends on the size of the organic ligand: high surface concentrations are more difficult to obtain with larger silane groups because of steric hindrance. Maximally-bonded packings will have the following concentrations (μmoles of bonded phase per m^2 of packing) for different silane groups: TMS (trimethylsilyl), 4.0–4.5; C-8, 3.0–3.3; C-18, 2.5–2.8. Poorer column-to-column retention-reproducibility and shorter column life can be expected for column packings with lower silane concentrations.

3.3 COLUMN PROBLEMS AND REMEDIES

Problems of one kind or another generally arise during the use of a column. This section considers how to recognize these problems and how to deal with them, so that effective method development is possible. We will now discuss the three most important kinds of column problems in HPLC method development: (a) retention and resolution reproducibility, (b) band tailing, and (c) column lifetime. Reference 9 should be consulted for routine procedures relating to the preventative maintenance and repair of columns.

Retention and Resolution Reproducibility

Reproducible retention and resolution for the peaks in a chromatogram are of major importance when developing routine methods that are to be used over a long time. Changes in resolution (arising from change in k', N, or α) can be a function of (a) the column and its operation, (b) instrumental effects, or (c) variations in separation conditions. Table 3.4 summarizes the types of retention and resolution variations that can occur in HPLC, and the causes for each variation.

Columns must maintain constant retention and resolution during use. Otherwise, the accuracy and precision of the method is compromised, and new columns may be required frequently. In some cases, a new column may give a different (unsatisfactory) separation; this may mean that the operating conditions for the method will have to be modified to reestablish the required separation. Often, the developed method must be transferred to another laboratory, where an equivalent column will be required for acceptable results. Therefore, the operator should be aware of the sources of column irreproducibility; practical remedies may be needed for handling this problem, as discussed below.

Retention reproducibility can be a major problem in developing a good HPLC method. Problems associated with irreproducibility can usually be solved by:

1. Selecting a column initially on the basis of its chemical nature, as shown in Table 3.1
2. Eliminating "chemical" or silanol effects by using various additives in the mobile phase to minimize band tailing (one symptom of silanol effects)
3. Utilizing proper laboratory techniques that ensure stable day-to-day operation
4. Using retention mapping to provide corrective action when required (Sect. 8.4)
5. Stockpiling columns, or establishing a continuing supply of the same column; alternatively, testing several column lots to ensure that the column lot selected will always work for the final method

TABLE 3.4   Retention and Resolution Variations in HPLC

Effect	Cause[a]
Initial differences in retention	Variations in silica support, bonding (k', α)
Column changes during use	Disturbance in bed (N); loss of bonded phase (k', α); dissolution of silica support (N); buildup of noneluted material (k', N)
Extra-column effects	From system-to-system: large injection volume; large tubing volume between injection valve and column and/or column and detector; large detector volume; large-volume fitting (N)
Poor control of separation	Changes in mobile phase, composition (k', α), flow rate (N), and temperature (k', α, N)
Slow column equilibration	Insufficient reequilibration time allowed (k', α)
Column overload	Too large a sample mass (k', N)

[a]k', N, and α refer to main changes in chromatogram.

As described earlier, the same type of column (e.g., C-18) from different manufacturers will often show substantial differences in both retention and resolution, due to variations in silica substrate and bonding chemistry. As a result, columns from different manufacturers may not be interchangeable. The results of separation on a C-18 column from company A will often differ from that obtained using a C-18 column from company B.

As discussed in Sect. 3.2 and illustrated in Fig. 3.4, column-to-column variations in retention can occur as a result of various chemical (silanol and other) effects; manufacturers currently are unable to guarantee that silanol effects will be constant from batch to batch of column packing. Broad, tailing peaks are an indication of major silanol effects and associated retention irreproducibility. Silanol effects usually can be minimized; the listing in Table 3.1 should be used to select an appropriate column when basic or acidic components are present in the sample of interest. However, in most reversed-phase separations, it is additionally desirable to create "generic" columns (minimizing silanol effects) by adding modifiers such as 150 mM acetate and 20–30 mM triethylamine to the mobile phase. Reference 18 contains a discussion of this approach in developing acceptable retention behavior for more than 150 drugs of pharmaceutical interest.

Poor retention reproducibilty and tailing peaks also can occur in separations that are poorly buffered, or where the ionic strength is too low. Increasing the buffer concentration (commensurate with sample size) can greatly improve this situation.

Changes in retention and resolution also can occur as the result of poor control of experimental conditions. Changes in the mobile phase can cause variations in the chromatogram, either during the day, or from day to day. Manually-prepared mobile phases should be carefully blended using solvents that are at the same temperature (weighing is the most accurate technique). Where possible, on-line mixing of the solvents by the instrument usually minimizes errors of this kind. However, when the mobile phase contains less than 10% of any one solvent (especially at low flow rates), on-line mixing can be *less* accurate than manual preparation of the mobile phase.

Variations in retention also can take place as a result of selective solvent fractionation by evaporation, either during degassing of the mobile phase, or on standing (however this problem is of minor significance in the case of reversed-phase HPLC; see Ref. 19). Solvent degassing, either by vacuum or preferably by helium purge (~5 min of vigorous sparging with a gas-dispersion tube) should be carried out by the same procedure each time, to ensure repeatability. Uptake of carbon dioxide, which can change the pH of the mobile phase, can be minimized by slowly and continuously bubbling helium though the mobile phase in the reservoir during use. If an error in mobile-phase composition is suspected, a new batch of mobile phase should be carefully prepared, and the separation repeated. Manually-blended solvents should be used to replace suspect on-line-mixed solvents.

Flow-rate variations can cause both sudden changes in the retention of all bands, as well as random fluctuations in peaks from run to run. Reference (9) should be consulted for the detection and solution of such problems.

Too large a sample can cause retention times and/or N-values to decrease for peaks that are overloading the column. The usual solution to this problem is to empirically determine the maximum sample size that gives constant retention times and plate numbers for peaks of interest. Typically, this sample size is 1–10 μg of sample per gram of column packing, depending on the system used. A useful approach is to start with a sample size in the upper range and then decrease the size until constant retention times are achieved.

A change in column temperature is a common cause of varying retention (see Fig. 5.7), especially when separating ionic or ionizable compounds (where significant variations in α can occur with temperature change). The column should be thermostatted to maintain the temperature to $\pm0.2°$ C. If this is not feasible, temperature variation can be minimized by reducing the change in temperature within the laboratory, or by insulating the column to reduce the effect of laboratory temperature changes. A constant temperature environment is particularly important when using automatic sampling for the analysis of a large number of samples. In this case, a change in column temperature will result in drifting retention times that may fall outside the narrow "windows" required by some automatic data handling systems.

Should changes in retention occur during the use of a method, predictable changes in the operating conditions (solvent strength, solvent mixture, etc.) often can be made so as to reestablish an acceptable separation. This goal can be realized without having to redevelop the method, by using retention "maps" for the compounds of interest; see Sect. 8.4.

Columns of a given type from the *same* manufacturer also can show significant retention and selectivity variations; a particular separation developed on one column may not be the same when a second column of the same type is used. A partial answer to this problem is to (a) use columns from a manufacturer who has the ability to deliver (and warrant) a high-quality, reproducible product, and (b) select columns from the same production lot. Some manufacturers can provide columns from the same (large) lot over a several-year period; such columns should be identical with respect to retention and chromatographic performance (e.g., the DuPont "Safeguard" program).

Another way to ensure that a method will function reliably with different columns is to evaluate columns from several different lots (batches). Acceptably reproducible retention and selectivity should be obtained for all lots, before the method can be considered trouble-free for routine application. Peaks of interest should be separated with at least the minimum resolution required for the desired measurement, taking loss in resolution with use into account.

Band Tailing

Conditions that result in tailing or asymmetrical peaks should be avoided. Band tailing generally results in inferior separations and reduced precision (especially when using automatic data systems); poor column-to-column reproducibility is also likely. In this section, we will discuss band tailing in method development as a function of the column and its history. Reference (9) should be consulted for a more general discussion of the problems and solutions associated with tailing bands.

Peak asymmetry or band tailing can arise from a number of sources, as summarized in Table 3.5. A "bad" column is a frequent source of asymmetrical peaks that can lead to poor quantitation. Asymmetrical peaks can arise from a column that was poorly packed. New columns showing undue peak asymmetry (Sect. 3.2) should not be used for method development, but returned to the manufacturer for replacement. During use, columns can develop severe band tailing (Fig. 3.9A) or even double peaks for each component (Fig. 3.9B). Such effects usually arise from either a partially-plugged inlet frit, or a void in the inlet of the column. Problems associated with a dirty inlet frit often can be eliminated by carefully replacing the inlet frit of the column (without disturbing the packing). Problems with voids at the column inlet sometimes can be reduced by filling the inlet with more packing (using a thick slurry); however, the original performance of the column is rarely restored by this approach (a possible exception is void-filling followed by reversing the flow through the column; see Ref. 20).

The development of broader, tailing peaks (as in Fig. 3.9) during the use of a column can signal the buildup on the column inlet of strongly retained "garbage" from the sample. This buildup sometimes can be eliminated by extensive purging of the column with very strong solvents (e.g., a 1/1 mixture of

TABLE 3.5   Causes of Asymmetrical (Tailing) Peaks

- "Bad" column
- Buildup of "garbage" on column inlet
- Sample overload
- Wrong solvent for sample
- Extra-column effects
- Chemical or secondary retention (silanol) effects
- Inadequate or inappropriate buffering
- Contaminating heavy metals

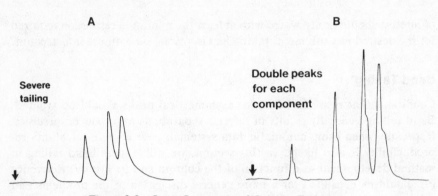

Figure 3.9   Some Symptoms of Column Problems.

isopropanol/methylene chloride for a reversed-phase column and methanol for a normal-phase column; in extreme cases, this purging should be carried out by back-flushing the column). However, in developing a routine method, the best procedure is to reduce the possibility of "garbage" buildup by using a guard column, as discussed in the next section. Alternatively, sample pre-treatment is also effective.

Overloading the column with sample can result in broadened, tailing (or fronting) peaks. This undesired effect usually can be eliminated by reducing the mass of sample injected (increase detector sensitivity, if required), until peak shape and retention stop changing. The resolution of the sample usually is also improved with smaller samples. Typically, 2–50 μg of each compound in the sample can be injected onto a 15 × 0.46 cm column without significant overloading.

Injecting the sample in a solvent stronger than the mobile phase usually results in distorted, tailing peaks, as illustrated in Figure 3.10. In this exam-ple, injecting the sample as a solution in pure acetonitrile produced broader, more skewed peaks, compared to a separation in which the sample was dis-solved in the acetonitrile/water mobile phase. When the sample is quite insol-uble in the mobile phase (or weaker solvents), small volumes (e.g., <25 μL for a 0.46 cm i.d. column) of sample in a stronger solvent can be injected. However, poorer peak shapes, sample precipitation, and perhaps column blockage, and/or compromised quantitation all may result.

Extra-column effects associated with the HPLC equipment can cause peak tailing and band broadening, as well as resulting poor quantitation. These band-spreading effects are associated with (a) relatively large volumes in the sampling valve, (b) long wide-diameter lines between the sampling valve, the column, and the detector, and (c) the volume of the detector flow cell. All such extra-column effects can combine to increase peak tailing. This type of

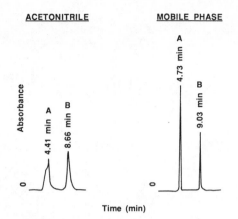

Figure 3.10   Sample-Solvent Effect. 30-μL sample volume; 18% acetonitrile/water mobile phase; caffeine (peak A) and salicylamide (peak B) injected in pure acetonitrile and in mobile phase. Reprinted with permission from Ref. (21).

tailing is most pronounced for early-eluting peaks, since they have the smallest volume (narrowest peaks). This effect is shown in Fig. 3.11a, where the early-eluting, narrower peaks from this column of 3-μm particles show significant tailing because of extra-column band broadening associated with this "standard" HPLC system. Later-eluting peaks of increasing volume exhibit progressively less tailing. This trend (early peaks tail most) is a good indication of the existence of extra-column effects in the apparatus. In Fig. 3.11b, peak tailing is less pronounced and retention times are shorter, because of the use of a lower-dead-volume "microbore" HPLC apparatus. Peak broadening and tailing due to extra-column effects should be eliminated or minimized before attempting to develop a separation. To do this, (a) inject small sample volumes (typically, < 25 μL); (b) use short connecting tubing of small internal diameter (e.g., < 20 cm of 0.010 in. i.d.) between the sample valve and the column, and between the column and the detector; (c) make sure that all tubing connections are made correctly from "matched" fittings; and, (d) use a cleanly swept, low-volume detector cell (< 10 μL). For useful discussions of extra-column effects, see Refs. 22 and 23.

Tailing or asymmetrical peaks also can occur because of various chemical effects, including a mismatch between the mobile/stationary phase combination and the sample (see Fig. 3.4). Such undesired effects can usually be eliminated by using mobile phases that contain acetate plus triethylamine (this is the "generic column" referred to in the previous section.) Sometimes, the problem with tailing peaks can only be removed by changing to an entirely

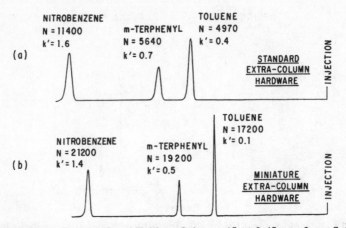

Figure 3.11 Extra-Column Band Tailing. Column, 15 × 0.45-cm, 3-μm Spherisorb silica; mobile phase, hexane/acetonitrile (99/1, v/v); flow rate, 2.0 mL/min. (a) commercial chromatograph with 10-μL sampling valve and 8-μL detector cell. (b) low-volume system with 0.5-μL sampling valve and 1-μL detector cell. Reprinted with permission from *Chromatographia*, Ref. (22).

different mobile phase/stationary phase combination (e.g., from reversed-phase to normal-phase).

Contaminating metals (Fe, Ni, etc.) in the column can produce band tailing for certain compounds. While metals can originate from the packing itself, they also can be deposited in the column packing by the slow dissolution of the metal frit in the column inlet by aggressive mobile phases. The tailing of basic drugs due to metal contamination of a C-18 packing is illustrated in Fig. 3.12; symmetrical peaks were obtained by acid-washing the packing before bonding.

Column Lifetime

Columns for normal-phase separations (especially bare silica) are often more stable than are columns used for the other HPLC procedures. It is not unusual for some normal-phase columns (e.g., silica, cyanopropyl) to have useful lifetimes of 1–3 yrs, when used with "clean" samples. Polymeric ion-exchange (resin) columns demonstrate similar stability. On the other hand, silica-based columns for reversed-phase, ion-pair, and ion-exchange chromatography are less rugged in the aqueous environments required for these separations. Even so, well-made columns can be stable under reversed-phase conditions for several months of continuous use (many hundreds or thousands of samples), if appropriate conditions are employed. Table 3.6 summarizes the

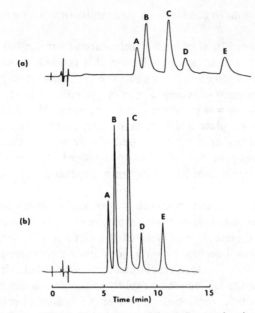

Figure 3.12 Tailing of Basic Drugs due to Metal Contamination of C-18 Silica. (a) Initial silica support, (b) after acid-washing the silica support; mobile phase, 20 mM TMA. Reprinted with permission from Ref. (24).

TABLE 3.6 Steps for Ensuring Good Column Lifetime and Performance

1. Use well-packed columns.
2. Minimize pressure surges; avoid mechanical and thermal shock.
3. Use a guard column and an in-line filter.
4. Frequently flush column with strong solvent.
5. Pretreat "dirty" samples to minimize strongly retained components of no interest and particulates.
6. Use a stable stationary phase.
7. Use column temperatures of $<60°$ C.
8. Keep the mobile-phase pH between 3 and 7 for silica-based columns; install a precolumn (saturator column) when operating outside these limits.
9. Add 100-ppm sodium azide to aqueous mobile phases and buffers.
10. For overnight and storage, purge out salt and buffers, leave column in pure acetonitrile, and avoid high concentrations of water and alcohols.

steps that can be taken to ensure good column lifetime and continued good performance.

Column lifetime can be significantly influenced by the initial condition of the packed column and by the way in which it is used. A sudden loss in column plate-number after relatively short use is usually caused by a sudden shifting of the packing, producing a void in the column inlet. This usually means that the column was poorly packed. Unfortunately, the *initial* condition of a column (i.e., plate number, asymmetry factor, etc.) often is not a good indicator of whether the column bed will be stable—this can only be determined in actual operation. A practical answer to this problem is to use columns from a supplier that has consistently produced stable columns.

Pressure Effects.   Sudden pressure surges and mechanical or thermal shock should be avoided. This will help minimize changes in peak shape or N-values that might dictate replacement of the column. Changes in mobile-phase flow rates should be made slowly, to minimize pressure pulses within the column. All types of sudden mechanical and thermal shock (e.g., dropping the column on the lab bench or rapidly changing column temperature) also should be avoided. Voids can also develop as result of pressure surges caused by slow valve actuation during sample introduction (this is a special problem in the use of some autosamplers). Losses in resolution from all of these sources can be minimized by using well-packed columns, and by operating at lower column pressures. However, it should be added that these pressure-related precautions are of minor importance as a means of extending column life; well-made HPLC columns are relatively rugged.

Sample "Garbage."   Column lifetime often can be significantly reduced by a buildup of strongly-sorbed sample components on the packing at the column inlet. This buildup of noneluted components is especially a problem with complex samples such as extracts of biological tissues or fluids (e.g., serum), oil-containing formulations, etc., but often is not a serious factor with relatively pure samples such as synthetic drugs. The buildup of sample "garbage" can be reduced by inserting a guard column between the sampling valve and the analytical column. The guard column, usually a well-packed, short length (< 5 cm) containing a packing equivalent to (or similar to) that in the analytical column, captures strongly retained sample components and prevents them from entering the analytical column. These guard columns should be replaced at regular intervals; that is, before the strongly retained components elute into the analytical column.

Longer column life is often enhanced by flushing the column frequently with a strong solvent (1/1 isopropanol/chloroform for reversed-phase; methanol for normal-phase), to remove strongly contained components that slowly build up on the column inlet. This cleansing of the column by strong solvents is conveniently accomplished in gradient elution runs by periodically allowing the strong ("B") solvent at the end of the gradient to purge through the column for at least 20–30 column volumes. Especially "dirty" samples should be pretreated to remove strongly retained components (late eluters), as well as particulates that might be present.

Particulates.   The injection of samples containing particulates will ultimately block the column inlet, thus reducing the normal lifetime of the column. Particulates also arise from the wear of the sample injector and pump seals. The use of a 0.5-μm in-line filter after the sample valve usually eliminates these problems. These low-volume filters are designed to minimize extra-column effects, and they can be easily replaced after a series of injections. In-line filters do not eliminate the desirability of initially removing obvious particulates from the sample by filtration or centrifugation prior to sample injection, since these particulates can also damage the sample valve.

Bonded-Phase Stability.   Column lifetime can be significantly affected by loss of the stationary phase during use. Stationary/mobile phase combinations that lead to a rapid loss of bonded phase and resulting changes in retention should be avoided (consult the column manufacturer's recommendations). It is well known that reversed-phase columns with short-chain-silane groups are relatively unstable in this respect. In highly aggressive mobile phases (e.g., 2.5 > pH > 7), some columns of this type can lose most of the organic phase within a few hours (4). Reversed-phase columns with long alkyl groups (C-8 or C-18) are usually considered to be relatively stable; however, even these columns show significant lose of bonded-phase when used at very low or high pH. This effect is shown in Fig. 3.13 for a C-8 column at pH 2.1, where the retention of a test solute is plotted vs. the volume of mobile phase that has passed through the column. In this case, retention decreases with time, due to loss of the bonded phase. In most applications, C-8 and C-18 columns show good long-term stability, provided that proper operating procedures (summarized in Table 3.6) are followed.

The stability of the bonded organic ligand on a reversed-phase column also is a function of the type and acidity of the silica used as the support (5,7). Packings made with fully-hydroxylated silicas having a homogeneous distribution of surface silanol groups demonstrate superior stability. Recent studies suggest that the stability of reversed-phase packings is a function of the pH of the silica surface (5). Better bonded-phase stability is often found for silica supports having a lower acidity (see Table 3.1).

Since the loss of stationary phase from silica-based columns is accelerated at higher temperatures, temperatures above about 60° C should be avoided.

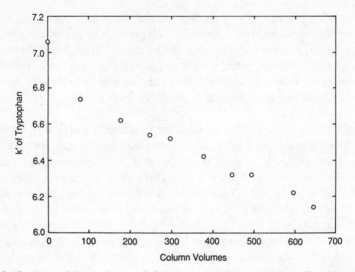

Figure 3.13   Loss of Retention on C-8 Column. k′ for tryptophan. Freshly prepared C-8 column, 25 × 0.46 cm; mobile phase, 40%v methanol/pH 2.1 phosphoric acid; 50° C; flow rate, 2.0 mL/min. Reprinted with permission from Ref. (25).

The insertion of a precolumn (saturation column) packed with silica prior to the sampling valve sometimes increases the stability of silica-based columns used under harsh operating conditions; e.g., $2.5 > pH > 7$ (26). This column (which can be packed with coarser particles) apparently saturates the incoming mobile phase with silica, thus retarding the dissolution of silica from the analytical column.

Microbial growth often occurs in buffers and aqueous mobile phases that are prepared and stored at ambient temperature for more than a day. Particulates from this source can plug the column inlet and significantly reduce column life. As a result, aqueous mobile phases that are free of organic solvents should be discarded at the end of each day. Alternatively, 100–200 ppm of sodium azide can be added to aqueous mobile phases to retard bacterial growth. Careful disposal of aqueous mobile phases containing azides (toxic and potentially explosive) is mandatory.

The performance and lifetime of reversed-phase columns are best preserved by storing unused columns in pure acetonitrile where possible (25). Storage with buffered solutions (particularly those containing high concentrations of water and alcohols) should be avoided. Columns should be capped tightly during storage, to prevent the column from drying out.

REFERENCES

1. L. R. Snyder and J. J. Kirkland, *Introduction to Modern Liquid Chromatography,* 2nd ed., John Wiley, New York, 1979, Chapt. 7.

2. K. K. Unger, *Porous Silica,* Elsevier, New York, 1979, Chapt. 2.

3. J. J. Kirkland, U.S. Patent 3,782,075, Jan. 1, 1974.

4. J. L. Glajch, J. J. Kirkland, and J. Köhler, *J. Chromatogr., 384* (1986) 81.

5. J. Köhler and J. J. Kirkland, *J. Chromatogr., 385* (1986) 125.

6. L. R. Snyder and M. A. Stadalius, in *High-Performance Liquid Chromatography. Advances and Perspectives, Vol. 4,* Cs. Horvath ed., Academic Press, New York, 1986, p. 195.

7. J. Köhler, D. B. Chase, R. D. Farlee, A. J. Vega, and J. J. Kirkland, *J. Chromatogr., 352* (1986) 275.

8. R. K. Iler, *The Chemistry of Silica,* John Wiley, New York, 1979, Chapt. 6.

9. J. W. Dolan and L. R. Snyder, *Troubleshooting HPLC Systems,* Humana Press, Clifton, N.J., 1989.

10. W. H. Pirkle and A. Tsipouras, *J. Chromatogr., 291* (1984) 291.

11. P. C. Sadek and P. W. Carr, *J. Chromatogr. Sci., 21* (1983) 314.

12. I. Wouters, S. Hendrickx, E. Roets, J. Hoogmartens, and H. Vanderhaeghe, *J. Chromatogr., 291* (1984) 59.

13. J. L. Glajch and J. J. Kirkland, U. S. Patents 4,705,725, Nov. 10, 1987; 4,746,572, May 24, 1988.

14. W. A. Moats, *J. Chromatogr., 366* (1986) 69.

15. J. Henion, in *Microcolumn Separation Methods,* M. Novotny, and D. Ishii, eds., Elsevier, Amsterdam, 1985, p. 243.

16. R. W. Stout, J. J. DeStefano, and L. R. Snyder, *J. Chromatogr., 282* (1983) 263.

17. J. J. Kirkland, W. W. Yau, H. J. Stoklosa, and C. H. Dilks, Jr., *J. Chromatogr. Sci., 15* (1977) 303.

18. R. W. Roos and C. A. Lau-Cam, *J. Chromatogr., 370* (1986) 403.

19. L. R. Snyder, *J. Chromatogr. Sci., 21,* (1983) 65.

20. J. Vendrell and F. X. Aviles, *J. Chromatogr., 356* (1986) 420.

21. J. W. Dolan, *LC-GC, 4* (1986) 16.

22. K. W. Freebairn and J. H. Knox, *Chromatographia, 19* (1984) 37.

23. J. W. Dolan, *LC-GC, 4* (1986) 1086.

24. M. Verzele, *LC Mag., 1* (1983) 217.

25. J. L. Glajch, J. C. Gluckman, J. G. Charikofsky, J. M. Minor, and J. J. Kirkland, *J. Chromatogr., 318* (1985) 23.

26. M. W. Dong, J. R. Gant, and P. A. Perrone, *LC-GC, 3* (1985) 786.

4

SYSTEMATIC METHOD DEVELOPMENT

This chapter will emphasize the initial steps in developing an HPLC method. Before actually injecting a sample, the first concern is to be sure that separated sample bands can be detected. In most cases, a UV (photometric) detector will be used if the sample components possess adequate ultraviolet (UV) absorptivity. In some cases, solute absorptivity will be marginal. The suitability of a UV detector must then be determined, together with the special steps that might be required to obtain the required detection sensitivity (see Sect. 4.1).

The main objective is getting an adequate separation with minimum time and effort. To assist in this objective, this chapter presents a systematic approach that is generally applicable to most samples; the scheme is outlined in

Table 4.1. The first step is to select an HPLC method and a column that seem suited to the particular sample. In most cases, it will be profitable to use one of three HPLC methods: reversed-phase, ion-pair or normal-phase chromatography; reversed-phase is by far the most commonly used method. Often, it will be possible to define all of the separation conditions by the end of Step 2 (Table 4.1); many samples can be separated by reversed-phase HPLC, with conditions similar to those listed in Table 1.2. For easily resolved samples, developing the separation should require only a few exploratory runs, and no more than a day in the laboratory (see Sect. 4.2).

More demanding samples require further development of the separation conditions. At this stage, this is best carried out by mapping mobile-phase selectivity for the particular HPLC method being used (Step 3, Table 4.1). In the case of reversed-phase HPLC, the choice of organic solvent (usually methanol, acetonitrile, or THF) to mix with water profoundly affects band spacing and resolution. The optimum exploitation of this approach requires both the selection of some minimum set of different solvents, and an efficient way to find the best mixture of this set of solvents (see Sect. 4.3).

A similar approach can be used for the other two major HPLC methods: ion-pair (Sect. 4.4) and normal-phase chromatography (Sect. 4.5). Various problems or special situations can arise from the approach given in Table 4.1, requiring minor modification of this scheme. Finally, information gathered during this systematic approach to HPLC method development can be used to further shape and improve the final choice of separation conditions.

Adjusting band spacing by changing mobile-phase selectivity should result in the satisfactory separation of >90% of all samples. When this approach is unsuccessful, the only alternative is to explore other ways of changing band spacing (Step 4 of Table 4.1). This option is detailed in Chapter 5.

TABLE 4.1   Systematic Approach to Obtaining an HPLC Separation

Step 1.	Choose an HPLC method and column that is suited for the sample in question.
Step 2.	Carry out two or more initial runs to determine how separation depends on solvent strength; choose a mobile-phase composition with respect to solvent strength.
Step 3.	If the separation resulting from Step 2 is inadequate, change the selectivity of the mobile phase (change in strong solvent, pH, additives, etc.), or try another HPLC method; map selectivity vs. mobile-phase composition for each HPLC method investigated so as to fully exploit its separation potential for the sample.
Step 4.	If further improvements in separation are required, systematically explore other variables: column type, temperature, etc. (Chap. 5).

4.1   CHOICE OF DETECTION CONDITIONS

In the early days of HPLC, only relatively crude, 254-nm photometers were available for detection. As a result, users often struggled to find ways to obtain adequate detection sensitivity. This situation has changed profoundly, as a result of the widespread use of commercially available, variable-wavelength spectrophotometric detectors with baseline noise of 0.0001 absorbance units (AU) or less. A low-wavelength (185–210 nm) UV detector usually can provide adequate sensitivity for most samples of interest. This section will discuss ways to achieve the full potential of UV detection.

Almost all unsaturated hydrocarbons and their derivatives possess a molar absorptivity ϵ of at least 10^3 in the wavelength range of 190–330 nm (see Table 4.2). This level of sensitivity is usually adequate for assays involving sample concentrations greater than 10 ng/mL. Many saturated aliphatic compounds with less-strongly absorbing functional groups (e.g., chloro, ester, amide, nitro, ether) have ϵ-values of >50 at appropriate wavelengths, as illustrated in Table 4.2 and Fig. 4.1. All of these compounds can be detected in the low-wavelength UV, provided the proper UV-transparent mobile phase is chosen. This situation is illustrated by the chromatograms of Fig. 4.2 ((a) 185 and (b) 195 nm; sample sizes of 0.4–600 μg). The main compounds for which a UV detector is not appropriate are saturated hydrocarbons, nitriles, and amines.

Therefore, when considering a UV detector for known compounds, the first step is to estimate the sensitivity (or determine the ϵ-values) for the sample compounds in the range of 190–330 nm. The appropriate wavelength can then be selected to allow the desired measurement of the compounds of interest. If a mixture of unknown compounds is to be detected, it is desirable in the first experiment to use the lowest wavelength that is practical. The wavelength can then be increased in succeeding experiments to meet various separation needs.

Minimum Detectable Mass

The minimum weight w_m of a compound (in μg) that will give a reasonable peak-maximum absorbance can be estimated from Beer's Law and various experimental conditions:

$$w_m = 1000M(k' + 1)(S/N)(N_0)L^{0.5}/\epsilon L_c d_c^2 N^{0.5}. \qquad (4.1)$$

Here, M is the molecular weight of the compound, k' is the capacity factor of the band, S/N is the required signal-to-noise ratio (usually >2), N_0 is the detector baseline noise in absorbance units, L is the column length (cm), L_c is the length (cm) of the detector flowcell, d_c is the internal diameter of the

TABLE 4.2 Representative Molar Absorptivity Values for Some Common Functional Groups

Compound Type	Chromophore	Wavelength (nm)	Molar Absorptivity (AU)
Acetylide	$-C{\equiv}C-$	175–180	6,000
Aldehyde	$-CHO$	210	1,500
		280–300	11–18
Amine	$-NH_2$	195	—
Azido	$C{=}N$	190	5,000
Azo	$-N{=}N$	285–400	3–25
Bromide	$-Br$	280	300
Carboxyl	$-COOH$	200–210	50–70
Bisulfide	$-S-S$	194	5,500
		255	400
Ester	$-COOR$	205	50
Ether	$-O-$	185	1,000
Iodide	$-I-$	260	400
Ketone	$C{=}O$	195	1,000
		270–285	15–30
Nitrate	$-ONO_2$	270	12
Nitrile	$-C{\equiv}N$	160	—
Nitrite	$-ONO$	220–230	1000–2000
		300–400	10
Nitro	$-NO_2$	210	strong
Nitroso	$-N{=}O$	302	100
Oxime	$-NOH$	190	5,000
Sulfone	$-SO_2$	180	—
Sulfoxide	$S{\rightarrow}O$	210	1,500
Thioether	$-S-O$	194	4,600
		215	1,600
Thioketone	$C{=}S$	205	strong
Thiol	$-SH$	195	1,400
Unsaturation, conjugated	$-(C{=}C)_3-$	260	35,000
	$-(C{=}C)_4-$	300	52,000
	$-(C{=}C)_5-$	330	118,000
Unsaturation, aliphatic	$-C{=}C-$	190	8,000
	$-(C{=}C)_2-$	210–230	21,000
Unsaturation, alicyclic	$-(C{=}C)_2-$	230–260	3000–8000
Miscellaneous compounds	$C{=}C-C{\equiv}C$	291	6,500
	$C{=}C-C{=}N$	220	23,000
	$C{=}C{=}C{=}O$	210–250	10,000–20,000
		300–350	weak
	$C{=}C-NO_2$	229	9,500
Benzene	C_6H_6	184	46,700
		202	6,900
		255	170
Diphenyl	$C_{12}H_{10}$	246	20,000

Source: Reprinted with permission from Ref. (1).

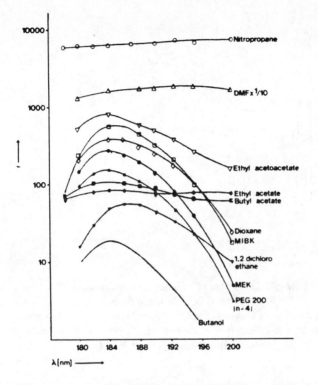

Figure 4.1   UV Spectra in the 180–200 nm Region for Various Aliphatic Molecules (Aqueous Solutions). Reprinted with permission from Ref. (2).

column (cm), and N is the column plate number. Table 4.3 shows two examples of the application of Eqn. 4.1. In Case A for unfavorable detection conditions, the minimum analyte mass must be at least 20 μg. In Case B for more favorable conditions, the minimum analyte mass is just 5 ng. The latter example represents about the best possible case under normal conditions. UV detectors can be used for quantitation at the low-nanogram level in selected cases.

Maximum Sample Size

The effect of sample size on bandwidth and retention can be predicted from the discussion in Ref. (4). For the 25-cm column used as example in the pre-

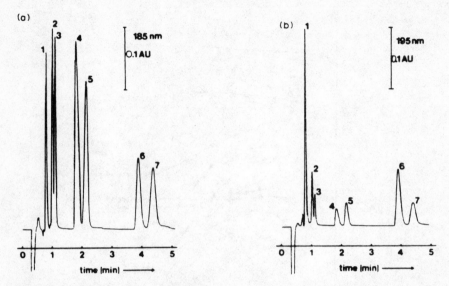

Figure 4.2 Low-UV Detection of Mixtures of Aliphatics. Reversed-phase separation. Column, 15 × 0.46-cm C-8; 7% acetonitrile/water. 1, Dimethylformamide (0.4 μg); 2, dioxane (17 μg); 3, tetraethylene glycol (34 μg); 4, butanol (650 μg); 5, methyl ethyl ketone (28 μg); 6, ethyl acetate (77 μg); 7, ethyl acetoacetate (16 μg). ketone (28 μg). Reprinted with permission from Ref. (2).

TABLE 4.3 Applications of Eqn. 4.1; Determining the Minimum Mass w_m of an Analyte for Quantitation[a]

	Detector Noise, N_0 (AU)	k'	ϵ	Minimum Mass, w_m (μg)
Case A	0.001	5	100	20
Case B	0.00005	2	10,000	0.005

FOR CASE A (FROM EQN. 4.1):

$$w_m = 1000M(k' + 1)(S/N)(N_0)L^{0.5}/\epsilon L_c d_c N^{0.5}$$
$$w_m = 1000(300)(6)(10)(0.001)(25^{0.5})/(100)(1)(0.46)(10000^{0.5})$$
$$w_m = 20 \ \mu g$$

[a]Molecular weight, 300; signal-to-noise ratio, 10; column, 25 × 0.46 cm; flow cell pathlength, 1 cm; N, 10,000.

vious section (Table 4.3), the maximum allowable sample size is about 100 μg under typical operating conditions. (A sample this large will result in about a 20% loss in N.) According to Table 4.3, a compound with $\epsilon = 100$ requires a sample size of 2–10 μg. Since about 10 times this amount of sample can be injected without significantly changing the separation, ϵ-values as small as 10 are potentially compatible with UV detection. Few workers appreciate the great versatility of a UV detector. For example (see Fig. 4.1), even aliphatic alcohols can be quantified with a UV detector operated at 185–190 nm.

However, for the use of UV detectors in the 185–205 nm range, care must be exercised in selecting mobile phases that are transparent in this wavelength region. Specially purified hexane or heptane are the nonpolar solvents of choice, and acetonitrile is the only polar organic solvent with acceptable absorbance. The choice of buffers for reversed-phase and ion-exchange chromatography at low wavelengths is somewhat limited; inorganic phosphates generally are the most suitable. Gradient elution is especially difficult in the 185–205 nm range, for several reasons (4). In special cases, it is feasible to establish conditions for minimizing baseline shifts during gradient runs by equalizing the absorbance of the two gradient solvents (A and B) with additives (2,4).

Maximizing UV Detectability

Two separate cases must be recognized: content analysis and trace analysis. For content assay, the compound of interest is the major component of the sample; large amounts of sample are available. If the sample has a weak UV chromophore, the objective is to inject as large a mass of sample as possible; for example, start with samples of about 100 μg. If the resulting peak is inconveniently large (off-scale), is in the nonlinear detection range, or if the separation is degraded by this large sample (100 μg), smaller samples can be tested.

In trace analysis (< 100 ppm) the compound of interest is present in relatively low concentration; the sample also may be available in limited quantity. These two factors limit the weight of the compound that can be injected, which may preclude UV detection for the actual compound of interest. However, a UV detector often can be used initially with higher-concentration mixtures of standard compounds to develop adequate separation conditions. This approach may be more convenient than using the actual detector required for the final HPLC method (with lower-concentration samples).

Finally, the baseline or background noise of the detector may be affected both by detection conditions (wavelength, filter band pass, temperature, mobile-phase fluctuations, etc.), and by separation conditions. Also, baseline noise under routine assay conditions can be an order-of-magnitude greater than initially observed without a column in the system (5). The main reason for the latter problem is the periodic elution of detectable material from the column during use (late eluters and/or column-packing degradation) that contributes to baseline variation. This effect can be reduced by:

(a) Washing the column daily with a strong solvent (e.g., methanol or acetonitrile)

(b) Pretreating the sample to eliminate late-eluters

(c) Using column-switching (6) or gradient elution (see Fig. 7.3b) to similarly remove late eluters in each run.

4.2   FIRST STEPS IN GETTING THE SEPARATION

The initial step in developing the separation is to select the most promising HPLC method and an appropriate column for the sample of interest. In most cases, the first separations should use reversed-phase HPLC with conditions similar to those of Table 1.2. Some exceptions to the initial choice of reversed-phase HPLC with a C-8 or C-18 column are summarized in Table 4.4. Table 1.3 offers reasons for considering ion-pair or normal-phase HPLC (instead of reversed-phase) for samples other than those listed in Table 4.2. We suggest

TABLE 4.4   Preferred HPLC Methods and Columns for Different Samples

Sample Characteristics	Preferred HPLC Method/Column
High molecular weight (>2000 daltons)	Special columns usually required; size-exclusion and ion-exchange HPLC often preferred (see Sect. 7.1, 7.2)
Optical isomers (enantiomers) present	Special chiral columns generally required (see Sect. 7.3)
Other isomers (stereo-, positional, etc.) present	Normal phase often best, especially with unmodified silica columns (see Sect. 5.4)
Mixtures of inorganic salts	Ion chromatography (see Sect. 7.1)
Carbohydrates	Amino-phase columns with reversed-phase conditions (see Sect. 7.2); ion-exchange resins
Biological samples	Special conditions often used for life-science samples; however, may not require different approach, compared with other low-molecular-weight compounds

that the reader study Table 4.4 and then Table 1.3 before choosing reversed-phase HPLC as the starting point for method development.

The following discussion assumes that reversed-phase HPLC plus the "standard" operating conditions of Table 1.2 have been selected. However, these remarks apply equally to the choice of one of the other primary methods of Table 1.3 (ion-pair, Sect. 4.4; normal-phase, Sect. 4.5). Table 1.2 suggests the use of an acetonitrile/water mobile phase for the initial separation. Many workers are reluctant to use acetonitrile instead of methanol, because acetonitrile is more expensive and is also thought to be more toxic. However, using this solvent represents only a 2–4% increase in the cost per sample. Acetonitrile is more toxic than methanol (7), but its toxic-threshold limit is only 10 times less than for rubbing alcohol! Other factors being equal, acetonitrile has major advantages over methanol: lower operating pressures, slightly higher solvent strength, and applicability for detection in the 185–205 nm range.

The major questions at this point are: (a) what percent acetonitrile is needed for a desired reversed-phase separation? and (b) what other mobile-phase additives are needed? No other additives are required for samples that do not contain acids, bases, or salts (e.g., alkyl sulfonates, quaternary ammonium compounds). Table 4.5 summarizes some recommended additives for other samples. The use of various additives to solve different separation problems was also discussed in Section 3.3.

TABLE 4.5   Recommended Additives for Reversed-Phase Mobile Phases

Sample Characteristics	Additive
Basic compounds (e.g., amines)	50 mM phosphate buffer, 30 mM triethylamine[a] (buffer to pH 3.0)
Acidic compounds (e.g., carboxylic acids)	50 mM phosphate buffer, 1%w acetic acid[b] (buffer to pH 3.0)
Mixture of acids and bases	50 mM phosphate buffer, 30 mM triethylamine[a], 1% acetic acid[b] (buffer to pH 3.0)
Cationic salts (e.g., tetralkyl quaternary ammonium compounds)	30 mM triethylamine[a], 50 mM sodium nitrate[b]
Anionic salts (e.g., alkyl sulfonates)	1% acetic acid[b], 50 mM sodium nitrate[b]

[a]Stronger amine additives such as dimethyloctylamine can be used, if required for good peak shape.
[b]Optional, may not be necessary.

Figure 4.3 is a good example of the need for optimizing the concentration of organic solvent in the mobile phase to obtain sample bands that have k'-values between 1 and 20. Mobile phases (a) and (b) (70% and 60% methanol) are much too strong, and the sample is poorly resolved. Mobile phase (c) (50% methanol) has $k' < 1$ for the first band, which would be difficult to quantify accurately because of the sloping baseline. Mobile phase (d) (40% methanol) provides adequate separation of the sample and is probably optimum. Mobile phase (e) (30% methanol) is too weak, and the last band has broadened to the point of poor detectability. The run time for this last separation is also excessive (60 min).

The dependency of k' on percent organic can be determined in systematic fashion (as in Fig. 4.3 or 4.4) by using a strong solvent for the first run and then successively reducing the percent organic until the right k' range is obtained. Usually, 20% increments in percent organic are convenient for approximating the desired solvent strength. The following k' ranges correspond to the chromatograms of Figs. 4.4a–d:

Percent Methanol	k' Range
100	0.1–0.3
80	0.6–1.7
60	1.8–8.5
40	7–50

From these values, it appears that mobile phases having 50–70% methanol will provide the desired range of $1 < k' < 20$, with 55% methanol being near optimum (Fig. 4.4e).

If resolution is adequate for a range of organic modifier concentrations with $1 < k' < 20$, then the strongest mobile phase usually is preferred. Run times will be shorter, and all peaks will be taller. At this point, if resolution is inadequate, the only remaining steps for improving sample separation are to change column conditions (Sect. 2.5), or to vary mobile-phase selectivity as discussed below.

Optimizing Mobile-Phase Selectivity

An adequate resolution of all components in the sample will generally not be observed for every mobile phase that provides a good k' range. An example is shown in Fig. 4.4c, for 60% methanol as mobile phase. In this case, although the sample contains 10 components, only nine distinct peaks can be seen.

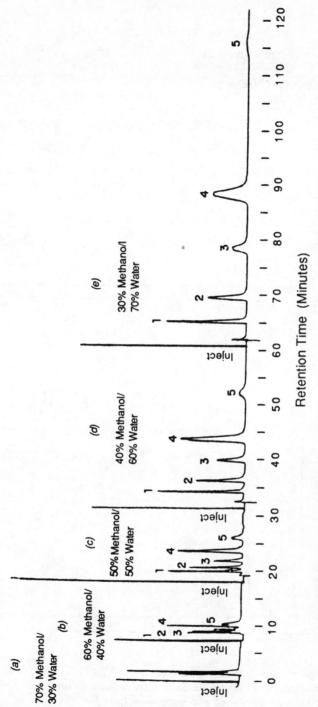

Figure 4.3 Reversed-Phase Separation of Alkylanthraquinone Mixture Using Mobile Phases of Varying Strength. Reprinted with permission from Ref. (8).

95

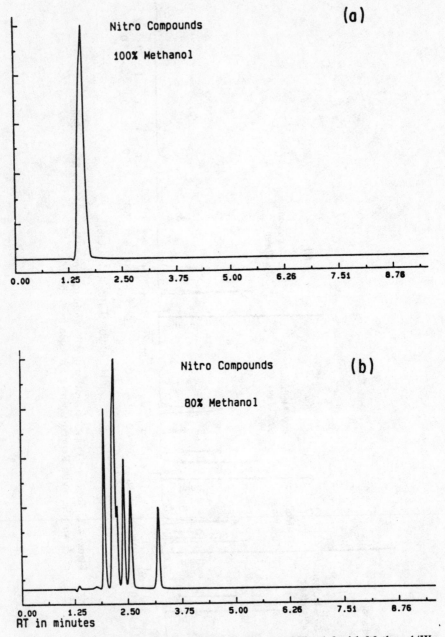

Figure 4.4 Separation of Nitro-Compound Mixture of Fig. 1.3 with Methanol/Water Mobile Phases. Arrows indicate unresolved band-pairs. Column, 15 × 0.46-cm Zorbax C-8, 2 mL/min. Reprinted with permission from Ref. (9).

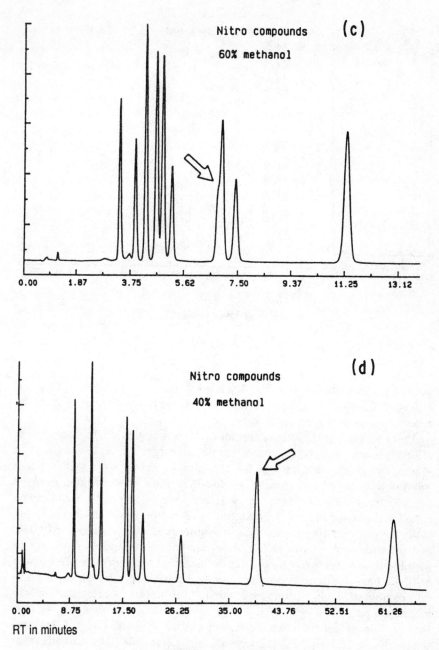

Figure 4.4 (*continued*)

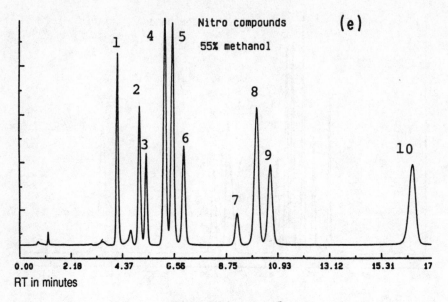

Figure 4.4 (*continued*)

Closer examination (arrow in Fig. 4.4c) shows that two bands are overlapped, with $R_s = 0.7$. If there is interest in either of these two components, it will be necessary to improve the separation.

There are three main alternatives for increasing resolution: (a) increase the column plate number N; (b) change the percent organic so as to change band spacing (while still keeping $1 < k' < 20$); and (c) change the mobile-phase solvents to change band spacing. (A fourth alternative—changing the type and selectivity of the stationary phase is less powerful; it can be used in some cases, as discussed in Sect. 5.2). Option (a) is not promising in this case, because the plate number for this 25-cm column is already in excess of 10,000, and the two bands (arrow in Fig. 4.4c) are badly overlapped. A change in mobile-phase solvents (e.g., acetonitrile instead of methanol) as described in Sect. 4.3 has considerable promise, but a systematic search for the right solvent combination can be time consuming. At this point, a better option is to investigate possible selectivity variations resulting from changes in k' vs. percent organic, rather than to use different organic solvents. Figure 4.4e shows the result of exploring different percent methanol mobile-phase mixtures that provide an adequate k' range (50–70% organic). Here, 55% methanol/water provides a good separation, and all 10 bands are resolved with $R_s > 1.3$. Only a minor improvement in the separation would be required to obtain an acceptable method. Such an improvement might be achieved by further exploring small changes in percent organic (e.g., 53%, 57%), or by optimizing column conditions (Sect. 2.5).

Preliminary data (10, 11) suggest that changes in band spacing as a result of change in percent organic in the mobile phase are relatively common. These changes in band spacing or α-values are more likely to be significant whenever the molecular sizes of two adjacent sample compounds are significantly different, or when their chemical functionality is dissimilar. However, even isomeric compounds can exhibit changes in band spacing as the percent organic is varied (e.g., Ref. 12). This means that the effect of percent organic on band spacing should be explored before trying more time-consuming options such as the use of different columns or solvents, change of pH or temperature, etc.

Gradient Elution for Initial Runs

An alternative to starting with 100% organic for the mobile phase in order to find the optimum solvent strength (as in Fig. 4.4a) is to use a preliminary gradient run. This approach was described in Sect. 2.2 (Fig. 2.13). There are a number of advantages to using a preliminary gradient run, compared to using isocratic runs as in Fig. 4.4. A single initial gradient run usually suffices to identify a mobile phase that has about the right solvent strength (see Sect. 9.3 and Fig. 9.6). This initial gradient run also provides significant separation of the sample components, in contrast to the initial isocratic result of Fig. 4.4a. Gradient runs are especially useful in the case of complex samples having a wide k' range; such samples often require gradient elution for the final method. Finally, gradient chromatograms generally have better front-end resolution than their isocratic counterparts, simplifying the interpretation of preliminary chromatograms by providing a larger number of identifiable bands.

Initial runs, either gradient or isocratic, may show that the k' range of the sample is excessive for a final isocratic procedure. That is, for any percent organic in the mobile phase, some sample k'-values always fall outside the range $1 < k' < 20$. In these cases, method development should be pursued as described in Chapter 6, with the goal of developing a gradient elution procedure for the final assay method. In a few cases, it has been found that, even when $1 < k' < 20$, only gradient elution is capable of achieving a satisfactory separation.

4.3 SOLVENT OPTIMIZATION: REVERSED-PHASE HPLC

The initial selection of the "right" HPLC method (e.g., reversed-phase), followed by the mapping of retention as a function of the percent strong solvent in the mobile phase, may not provide acceptable resolution of the sample. These "difficult" samples can require considerably more time and effort to develop a satisfactory final separation: perhaps a week or more for achieving separation, as opposed to one or two days using the approach of Sect. 4.2. However, it is possible (and desirable) to proceed so that the easier samples in this more-difficult group are handled as efficiently as possible, thus minimizing the time required for method development. This section presents such a strategy.

Basis of Solvent Selectivity

In optimizing band spacing by changing the organic solvent (e.g., methanol, acetonitrile, and propanol), a compromise must be reached between maximum selectivity and a minimum number of required experiments. It is desirable if a small number of solvents can provide a wide range of selectivity changes. As discussed in Sect. 2.3, three solvents appear to meet these requirements for reversed-phase HPLC: methanol (MeOH), acetonitrile (ACN), and tetrahydrofuran (THF). Figure 2.15 also shows that various blends of these solvents with water can simultaneously control both the k' range and band spacing, so that an optimum resolution of sample components can be achieved in systematic fashion.

Three different approaches to systematic mobile-phase mapping and optimization in reversed-phase HPLC will now be examined. These approaches, and their relationships to preceding steps in method development (Sect. 4.2), are diagrammed in Fig. 4.5. In this figure, samples are arbitrarily classified as "easy," "average," and "hard." "Easy" samples are those that are successfully separated as described in Sect. 4.2, using only variations in percent acetonitrile to simultaneously achieve a good k' range and adequate selectivity (band spacing). "Average" samples are those that are not adequately resolved in this way, but which can be separated using a combination of (a) varying the percent organic and (b) using methanol or THF instead of acetonitrile in the mobile phase. "Hard" samples require complete solvent optimization, where resolution is mapped as a function of varying proportions of acetonitrile, methanol, and THF in the mobile phase. Usually, the k' range (or run time) is held constant during this latter process, by varying the amount of water in the mobile phase mixture so as to compensate for small differences in the strength of the three pure organic solvents.

"EASY" SAMPLES

| Initial runs with ACN/water | → | Define k'-range vs %-ACN | → | Best %-ACN for k'-range and band spacing |

"AVERAGE" SAMPLES

| Initial runs with MeOH/water | → | Define k'-range vs %-MeOH | → | Best %-THF for k'-range and band spacing |

| Initial runs with THF/water | → | Define k'-range vs %-THF | → | Best %-MeOH for k'-range and band spacing |

"HARD" SAMPLES

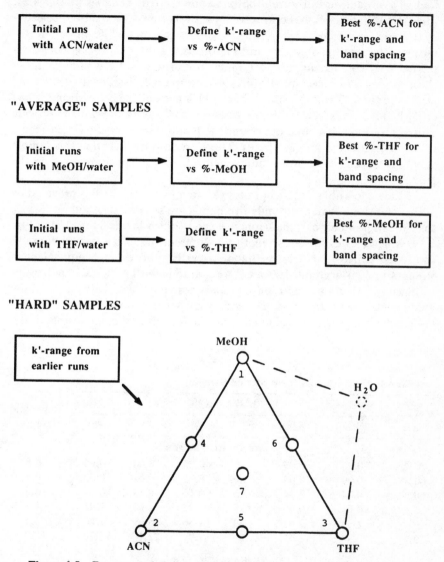

Figure 4.5 Recommended Approach for Developing an HPLC Separation.

"Average" Samples.   The approach for "average" samples can be visualized as a variant on full-solvent optimization ("hard" samples in Fig. 4.5). The main difference is that only solvents at the corners of the solvent-selectivity triangle are studied, that is, only the binary-solvent mobile phases: acetonitrile/water, methanol/water, and THF/water. Table 4.6 provides an example for a six-component steroid sample. Initial runs are the same as those chosen for easy samples: mobile phases of acetonitrile/water with decreasing percent organic. Table 4.6 shows that $\geq 40\%$ acetonitrile is too strong; some or all bands have $k' < 1$. Mobile phases with 20–30% acetonitrile provide satisfactory retention, but compounds # 1 and 2 are not separated (same k'-values) with both mobile phases. Therefore, it is unlikely that this sample could be separated by simple adjustment of the percent acetonitrile in the mobile phase.

As shown in Table 4.6, an optimum k' range ($2 < k' < 8$) is provided by 30% acetonitrile/water, but only five bands are separated. Another organic solvent can now be tried; methanol is a good choice at this point. Using Fig. 2.14, which relates solvent strength for different organic/water mixtures, it can be estimated that 40% methanol/water is approximately equivalent in strength to 30% acetonitrile/water. A separation with 40% methanol/water is shown in Table 4.6; well-resolved bands were obtained. Adjustment of the methanol concentration was explored to further improve band spacing, but 40% methanol/water proved to be optimum for this particular sample. If de-

TABLE 4.6   Retention Data for Steroid Sample

Mobile Phase	k'-Values						Run Time (min)
	Peak: #1	#2	#3	#4	#5	#6	
	ACETONITRILE/WATER						
100%	0.00	0.00	0.00	0.00	0.00	0.00	0.5
80	0.00	0.00	0.00	0.01	0.02	0.02	0.5
60	0.04	0.04	0.05	0.10	0.19	0.23	0.6
40	0.66	0.66	0.75	1.32	2.03	2.41	1.5
30	2.63	2.63	2.89	4.86	6.69	7.86	4.4
20	10.4	10.4	11.2	17.9	22.1	25.7	13.3
	METHANOL/WATER						
40%	5.49	6.02	6.66	11.0	13.5	16.4	8.2

[a]Compounds: prednisone (#1), hydrocortisone (#2), cortisone (#3), dexamethasone (#4), corticosterone (#5), cortexolone (#6). Conditions: 8 × 0.62-cm, 3-μm, Zorbax C-8; mobile phases as shown, 3 mL/min; 35° C.
Source: Reprinted with permission from Ref. (9).

sired, a further increase in resolution could be sought by optimizing column conditions (see Sect. 2.5).

"Hard" Samples. The concept of systematically optimizing the proportions of methanol, acetonitrile, THF, and water in the mobile phase by mapping resolution vs. mobile-phase composition was first reported by Glajch et al. (13). This approach will be illustrated for the same steroid sample used in the preceding example (but with a different column geometry). The first step was the same as in Table 4.6: obtaining the proper percent organic for a reasonable k' range. In this case, starting with methanol/water mobile phases, it was determined that 49% methanol provided a good k' range. The chromatogram with this solvent is shown in Fig. 4.6 (mobile phase 1, at the top of the triangle).

Acetonitrile was the next choice for changing band spacing. Reference to the solvent-strength diagram of Fig. 2.14 suggested that 40% acetonitrile should be approximately equivalent to 49% methanol. However, this mobile phase proved to be too strong (run time of only 2 min, vs. 7 min in Fig. 4.6, mobile phase #1). In this case, use of Fig. 2.14 missed the mark slightly. Thus, the solvent-strength nomograph of Fig. 2.14 should be considered as accurate to only about ±5%; it can be in error by ±10% (14). Solvent strength can be adjusted, using the rule of thumb that a 10% decrease in organic will produce a threefold increase in run time. This rule suggests 30% acetonitrile for the steroid sample, instead of the 40% determined from the nomograph. The chromatogram with 30% acetonitrile is shown in Fig. 4.6 (mobile phase 2); the run time is now satisfactory.

Finally, Fig. 2.14 suggests that 30% acetonitrile is equivalent to 22% THF, and the latter run is shown in Fig. 4.6 (mobile phase 3). Now, the run times for all three mobile phases (1, 2, and 3 of Fig. 4.6) are about equal (~7 min). However, none of these three mobile phases yields adequate resolution of all six sample components.

The three equal-strength mobile phases (1, 2, and 3 of Fig. 4.6) were next blended to obtain mobile phases 4, 5, 6, and 7 of Fig. 4.6:

Volumes of Binary Mobile Phases[a]

Mobile Phase	MeOH/Water	ACN/Water	THF/Water
4	1	1	0
5	0	1	1
6	1	0	1
7	1	1	1

[a]49% MeOH/water, 30% ACN/water, and 25% THF/water.

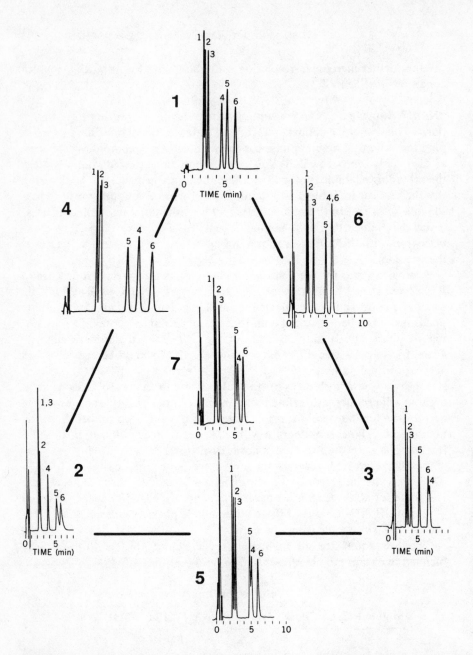

Figure 4.6 Solvent Optimization for the Reversed-Phase Separation of Six Steroids. Similar conditions as those in Table 4.5. Column, 15 × 0.46-cm Zorbax C-8. Mobile phases as in Fig. 4.5: 1, 49% methanol/water; 2, 30% acetonitrile/water; 3, 22% THF/water; other mobile phases are blends of mobile phases 1-3. Reprinted with permission from Ref. (8).

The resulting chromatograms for all seven mobile phases discussed above are shown in Fig. 4.6. Significant change in band spacing occur as the proportions of the different organic solvents are varied. Specifically, runs 5 and 3 show that good resolution of bands 1–3 is obtained in this region of the solvent triangle; band 4 also moves significantly as the ratio of ACN/THF is varied. This effect suggests varying the ACN/THF ratio by blending mobile phases 5 and 3, to place band 4 equidistant between bands 5 and 6. The result of blending these two mobile phases is shown in Fig. 4.7; chromatograms are given for both a 15-cm, 6-μm column (a) and an 8-cm, 3-μm column (b). These two chromatograms show the further improvement that is possible by optimizing column conditions (shorter run time in Fig. 4.7b).

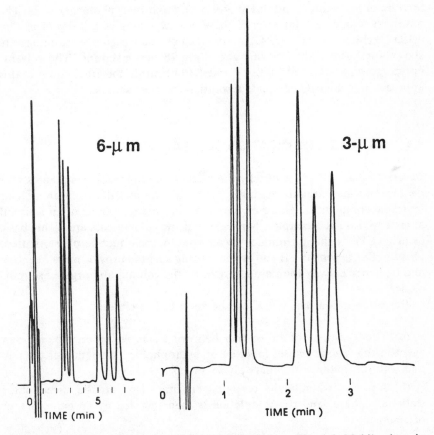

Figure 4.7 Optimum Separation of Steroid Sample from Fig. 4.6. Mobile phase is 11% acetonitrile, 12% THF, and water. Columns are 15 × 0.46-cm, 6-μm, or 8 × 0.62-cm, 3-μm. Reprinted with permission from Ref. (15).

As illustrated in Figs. 4.6 and 4.7, solvent optimization is a powerful technique for getting the best final separation. However, this approach can require 10 or more runs to arrive at an optimum separation (as in Fig. 4.7). Even more effort is required if standards are run to identify every band in each separation. So, this approach should be reserved for well-defined "hard" samples, those that are not successfully separated by the alternative procedures of Fig. 4.5. However, it is important to note that whatever conditions are used for separating a sample, the scheme of Fig. 4.5 does not waste any runs. All the information obtained from the easy-sample approach is applicable for the average-sample procedure. In turn, all of this information is applicable to the hard-sample scheme of Fig. 4.6.

Additional examples of solvent optimization are shown in Figure 4.8 for a mixture of herbicides, and in Fig. 4.9 for a mixture of photochemicals. The optimum separations for each of these last two samples are shown in Fig. 4.10a (herbicides) and Fig. 4.10b (photochemicals). Again, separations are shown for both 6-μm (15 cm) and 3-μm (8 cm) columns. The solvent-optimized separations of Figs. 4.7 and 4.10 illustrate the great power of this approach in achieving adequate resolution for most samples.

4.4   SOLVENT OPTIMIZATION: ION-PAIR HPLC

Reversed-phase HPLC is usually the method of choice for compounds that dissolve in water/organic mixtures, even with samples that contain ionic or ionizable compounds. However, ion-pair chromatography is often a useful alternative for such samples, particularly if the components are highly basic or acidic. With this technique, an aqueous/organic mobile phase is used, plus a buffer to control pH and an ion-pairing agent to provide more retention and higher selectivity than are afforded by the column and organic solvents alone.

Retention in ion-pair HPLC can be described as either:

(a) The formation of an ion-pair between a sample ion and the ion-pair agent in the mobile phase, followed by hydrophobic association of the ion-pair with the column stationary phase, or,

(b) A process whereby the ion-pair agent first adsorbs to the surface of the stationary phase, and the sample ion is then retained by an ion-exchange mechanism.

There is considerable support in the literature for either possibility; one or the other can be used to explain ion-pair chromatography. The discussion here

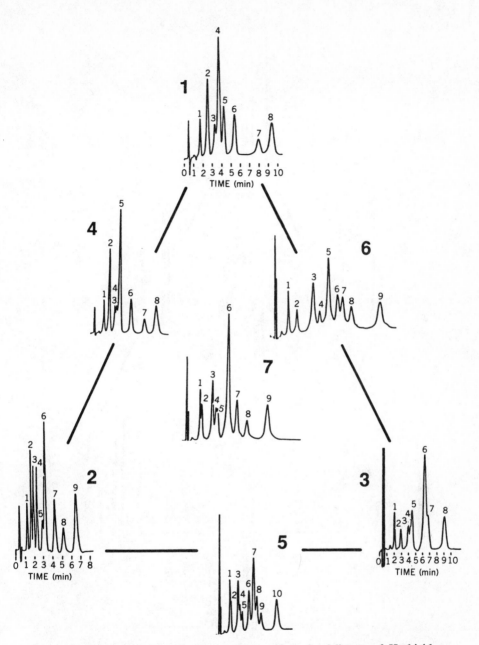

Figure 4.8 Solvent Optimization for the Separation of a Mixture of Herbicides. Column, 15 × 0.46-cm Zorbax ODS (C-18); 3 mL/min of each mobile phase (#1, 60% methanol; #2, 51% acetonitrile; #3, 36% THF); 50° C; compounds are (1) 2,4-D; (2) Ramrod; (3) 2,4,5-T; (4) 2,4-D methyl ester; (5) CIPC; (6) 2,4-DB; (7) Silvex; (8) 2,4,5-T-methyl ester; (9) 2,4-DB methyl ester; (10) Silvex methyl ester. Reprinted with permission from Ref. (8).

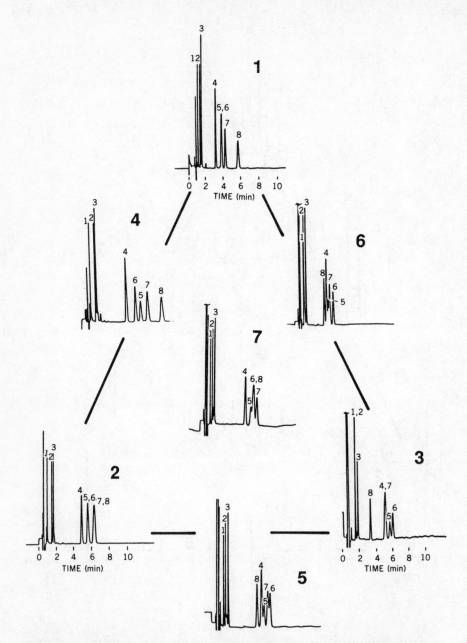

Figure 4.9 Solvent Optimization for the Separation of a Mixture of Photochemicals. Column, 15 × 0.46-cm Zorbax C-8; 3 mL/min of each mobile phase (#1, 62% methanol; #2, 43% acetonitrile; #3, 32% THF); 50° C; compounds are: p-methoxyphenol, diethylene glycol diacrylate, acetophenone, benzophenone, cyclohexylacrylate, 1,6-hexanedioldiacrylate, dimethoxyacetophenone, Michler's ketone. Reprinted with permission from Ref. (8).

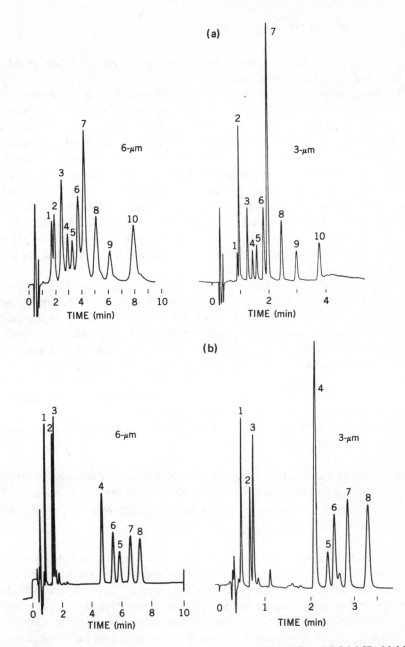

Figure 4.10 Optimum Separations of Samples from Figs. 4.8 and 4.9 (a) Herbicides, 9% methanol, 32% acetonitrile, and 9% THF, plus 1% acetic acid (for ion suppression); (b) photochemicals, 22% methanol, 26% acetonitrile, 1% THF. Reprinted with permission from Ref. (15).

will focus on the latter hypothesis. A good review of ion-pair chromatography can be found in Ref. (16, 17).

The ion-pair retention of an acidic compound HA is illustrated in Fig. 4.11. If no ion-pair agent is present (Fig. 4.11a), retention is controlled by reversed-phase partitioning. The molecule, HA, will exhibit a maximum k'-value (HA not ionized) when pH $\ll$ pK_a for the acid. At higher pH-values, the acidic molecule will be ionized, and retention will decrease significantly (see also Sect. 5.2). If a large concentration of ion-pair agent is present, the surface of the stationary phase will be essentially covered with the ion-pair agent. The ion-pair process then dominates retention, as illustrated in Fig. 4.11b. At low pH, the nonionized HA will not be retained by the ion-exchange process (but may be weakly retained by reversed-phase interaction). As pH is increased, HA ionizes to form the ion A^-. This species is retained via ion exchange with the sorbed ion-pair agent.

This example of ion-pair retention (Fig. 4.11) assumes that either no ion-pair agent is present (4.11a), or enough is present to completely convert the retention process to ion-exchange (4.11b). Intermediate levels of the ion-pair agent will result in a mixed process with retention occurring by both reversed-phase and ion-pair chromatography. Therefore, the concentration of ion-pair agent is very important in determining selectivity effects for band spacing. The type of ion-pair agent (e.g., C-5 vs. C-7 sulfonate) also alters band spacing. However, the effect of this variable is often similar to that obtained by changing the concentration of the ion-pair agent (16, 17).

The preceding discussion identifies the primary selectivity effects in ion-pair chromatography as:

(a) A change in retention from reversed-phase to ion exchange (or vice-versa) by varying the concentration of ion-pair agent in the mobile phase; and

(b) A change in retention of compounds with different pK_a-values, by varying mobile phase pH.

Since reversed-phase and ion-exchange chromatography operate by separate processes, different compounds can show changes in relative retention (α) as the concentration of ion-pair agent is changed. Ionizable compounds often have different pK_a-values and will show corresponding changes in retention as pH is changed. Therefore, the ion-pair agent concentration and mobile-phase pH are the most important variables for adjusting band spacing in ion-pair HPLC.

The general approach for changing band spacing in reversed-phase HPLC (Sect. 4.3), can be altered slightly so as to apply to ion-pair HPLC. "Easy" and "average" samples can be handled by approaches that are parallel to

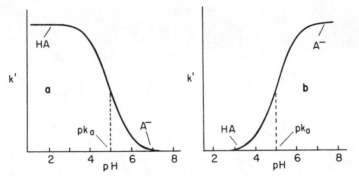

Figure 4.11   Retention of an Acidic Sample HA as a Function of pH. (a) Reversed-phase chromatography; (b) ion-pair chromatography with a large concentration of the ion-pair agent.

those described for reversed-phase chromatography. "Hard" samples require an optimization procedure similar to that for solvent optimization. However, as illustrated in Fig. 4.12, only one organic solvent (methanol) is used in optimizing separations in ion-pair chromatography; the other three components of the mobile phase are buffers of pH = 2.5 (B), pH = 7.5 (C), and another buffer of pH = 5.5 containing the maximum concentration (e.g., 200 mM) of the ion-pair agent (D). Seven experiments are run using different combinations of the buffers (B–D), with the amount of methanol (A) varied to maintain the k' range for all runs approximately constant (See Sect. 9.5).

An example of this approach is shown in Fig. 4.13 for a mixture of five compounds from a cough/cold remedy using a C-8 column and 200 mM hexanesulfonate as the ion-pair agent (18). Examination of chromatograms 1–3 in Fig. 4.13 (the corners of the triangle corresponding to binary mixtures of methanol and one of the three buffers) shows that band spacing is affected by changes in both pH and ion-pair concentration. This effect suggests that some intermediate mobile phase will optimize the separation of all five of these compounds.

Runs 4–7 in Fig. 4.13 are the chromatograms from ternary and quaternary mixtures, similar to those in Sect. 4.4 (Figs. 4.6, 4.8, and 4.9) for the reversed-phase solvent optimization procedure. Both runs 4 and 5 provide separation of all compounds, but in each case one solute elutes near t_0. Run 7 provides a more reasonable retention ($k' = 1$) for the first peak, suggesting that a mixture of mobile phases 4 and 7 can provide a better separation, as shown in Fig. 4.14. All five compounds are now well resolved and are also separated from t_0. Manual examination of chromatograms (e.g., Figs. 4.6–

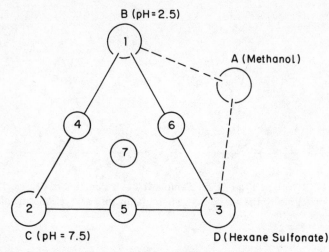

Figure 4.12   Systematic Variation of Mobile Phase for Optimizing Band Spacing in Ion-Pair HPLC. Reprinted with permission from Ref. (18).

4.10) often provides satisfactory results; a more precise version of this approach is discussed in Chapter 8, where appropriate computer software is used to make final predictions.

The previous approach (Fig. 4.12) used citric acid buffers (18), which provide good pH-control over the range $2 < pH < 8$. However, citrate buffers have a practical limitation in that citrate attacks the stainless steel components of most HPLC pumps and tubing. Phosphate buffers are an alternative, but the range is limited to $pH = 1.8-3.5$ and $5.8-8.0$. Acetate buffers are another option, but are limited to $pH = 3.7-5.6$. For method development we recommend use of a mixed buffer system based on acetic and phosphoric acids (pH adjusted with NaOH); these solutions provide good buffering over the range $pH = 2.5-7.5$.

The preceding discussion assumes that the ion-pair separation of a *basic* sample is desired, and that hexane sulfonate is selected as the ion-pair agent. Other ion-pair agents can be used, depending on the nature of the sample (see Ref. 16). When acidic compounds are to be separated, it is necessary to use a cationic ion-pair agent such as tetrabutylammonium ion (TBA). Detailed procedures for the ion-pair separation of both acidic and basic samples are given in Section 9.5. There are many other ion-pairing agents to choose from; e.g., C-5–C-12 sulfonates, C-1 and higher tetralkylammonium ions, etc. Retention and band spacing in ion-pair HPLC are determined mainly by the total amount (moles) of ion-pair agent taken up by the column packing, and

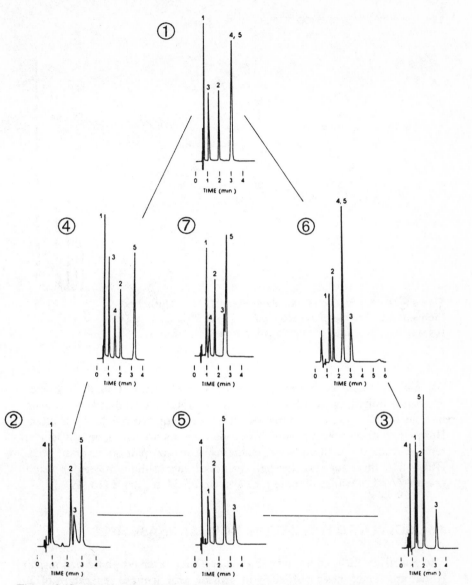

Figure 4.13 Application of Method of Fig. 4.12 to the Separation of a Mixture of Compounds in a Cough/Cold Remedy Separation on a C-8 column; mobile phases contain:

Solvent	%v Solvent in Mobile Phases #1-7						
	1	2	3	4	5	6	7
A (methanol)	30	27	34	29	30	32	30
B (pH = 2.5)	70	0	0	35	0	35	23
C (pH = 7.5)	0	73	0	36	36	0	24
D (ion-pair)	0	0	66	0	33	33	22

Compounds are: 1 = phenylephrine, 2 = glycerol guaicolate, 3 = pseudoephedrine, 4 = sodium benzoate, 5 = methylparaben. Reprinted with permission from Ref. (18).

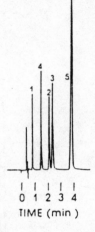

Figure 4.14   Separation of Sample of Fig. 4.13 with Optimized Mobile Phase. Mobile phase is 30% A, 25% B, 28% C, 18% D (by volume). Reprinted with permission from Ref. (18).

are less dependent on the exact ion-pair agent selected. Usually, a higher-molecular-weight agent will be more strongly retained, so that lower concentrations of the agent (in the mobile phase) are required for the same effect. However, higher-molecular-weight ion-pair agents are also more difficult to remove from the column when changing from ion-pair to reversed-phase HPLC. For these reasons, ion-pair agents of intermediate molecular weight are preferred: hexanesulfonate for bases and TBA for acids (18, 19).

4.5   SOLVENT OPTIMIZATION: NORMAL-PHASE HPLC

The selectivity differences available from either reversed-phase or ion-pair HPLC are sometimes insufficient to provide an adequate separation of all bands of interest. A change to normal-phase HPLC can be useful in these cases, since different retention processes provide different selectivity effects. In addition, the solubility of many organic compounds is often greater in normal-phase HPLC solvents. This factor can be a strong advantage if preparative HPLC is a final separation goal.

Until recently, normal-phase HPLC implied the use of unmodified adsorbents, such as silica or alumina. Although selectivity effects can be impressive on these supports (most isomers are routinely separated by normal-phase silica chromatography with baseline resolution), certain practical considerations have limited the use of these unmodified adsorbents for HPLC analysis. Foremost among these is the need to carefully control the water content of the mobile phase to obtain reproducible results (20). Also, it is impractical to use unmodified adsorbents for reproducible gradient-elution separations, because of difficulties in maintaining equilibrium between the adsorbent and the mobile phase with resulting retention irreproducibility and slow reequilibration of the column after each gradient run.

Fortunately, recent work (e.g., Refs. 21 and 22) has shown that certain bonded-phases (cyano, diol, and amino) are more useful and convenient substitutes for bare silica. These packings can be used quite satisfactorily for gradient elution. We recommend, therefore, that method development in normal-phase HPLC begin with one of these bonded phases; cyano is usually a good first choice. An unmodified adsorbent (usually silica) should only be used when other attempts have failed to provide an adequate separation, or if preparative HPLC is a final goal (23).

In normal-phase HPLC, the sample retention is governed by adsorption to the stationary phase. For retention to occur, a sample molecule must displace one or more solvent molecules from the stationary phase, as shown in Fig. 4.15. In this case, a molecule of sample S displaces two molecules of solvent E. In addition to this *displacement* effect, polar solvent or sample molecules can exhibit very strong interaction with particular sites on the stationary phase. This effect, termed *localization,* is shown in Fig. 4.15b, where polar substituents (X, Y) on a benzene ring interact more strongly with the adsorption sites A on the surface than does the benzene ring itself.

These two effects, displacement and localization, are the primary sources of mobile-phase selectivity in normal-phase HPLC. The primary mobile-phase properties which control solvent selectivity are summarized in Fig. 4.16. This classification is based on a nonpolar solvent as the diluent (analogous to water in reversed-phase), plus three polar organic solvents (nonlocalizing, basic localizing, and nonbasic localizing) to vary band spacing for the sample. Typically, hexane or FC-113 (1,1,2-trifluoro-1,2,2-trichloroethane) (24) is used as the nonpolar solvent, and the three modifiers can be chosen (one each) from the groups in Table 4.7 (25). Hexane is useful for separations involving low UV detection, but it is not totally miscible with all of the solvents in Table 4.7. Therefore, when using hexane, co-solvents (e.g., methylene chloride) may have to be added to the mobile phase to ensure miscibility. In contrast, FC-113 is miscible with all the solvents in Table 4.7, and provides the maximum available selectivity. While exhibiting significant UV absorbance below 230 nm, FC-113 is the nonpolar solvent of choice for most normal-

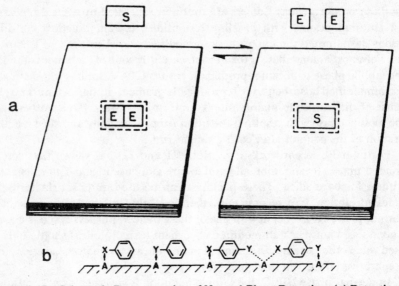

Figure 4.15 Schematic Representation of Normal-Phase Retention. (a) Retention of sample molecule S with displacement of solvent molecules E; (b) Localization of substituted-benzene samples on adsorption sites A. Reprinted with permission from Ref. (20).

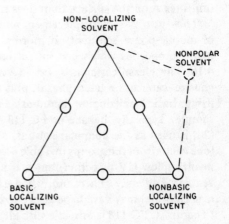

Figure 4.16 Experimental Design for Solvent Optimization in Normal-Phase HPLC. Solid circles indicate solvent mixtures for scouting runs. Reprinted with permission from Ref. (28).

TABLE 4.7   Classification of Solvents According to Band-Spacing Effects in Normal-Phase HPLC

Weakly Polar (Nonlocalizing)	Polar (Localizing)	
	Basic	Nonbasic
Methylene chloride[a]	Methyl-t-butyl	Acetonitrile[a]
Chloroform[a]	ether (MTBE)[a]	Ethyl acetate[a]
Toluene	Ethyl ether	Acetone
Chlorobutane[a]	Isopropyl ether	Nitromethane
Ethylene chloride[a]	Tetrahydrofuran	
	Dioxane	
	Triethylamine	
	Dimethylsulfoxide	
	Methanol, propanol[a,b]	

[a]Preferred solvents (do not adsorb above 240 nm and are relatively stable).
[b]These solvents have added proton-donor selectivity.
Source: Reprinted with permission from Ref. (25).

phase applications. Mixtures of hexane, methyl-t-butyl ether, and acetonitrile are also available for use below 230 nm, although with a reduced selectivity range.

Adjustment of Solvent Strength

The first priority in solvent optimization for normal-phase HPLC is to obtain the proper solvent strength. All compounds should elute in the range of $1 < k' < 20$ as described in Sect. 2.2. In normal-phase HPLC, solvent strength can be described by the parameter $\epsilon°$ for silica, alumina, or bonded-phase columns. The calculation of $\epsilon°$ for a particular solvent mixture has been summarized (26), but is somewhat complex. Tables 2.2 and 4.8 provide convenient data for $\epsilon°$-values on silica and bonded-phase columns, respectively. For example, if a solvent strength of $\epsilon° = 0.25$ were required on a silica column, Table 2.2 suggests that this solvent-strength value can be obtained with 58% methylene chloride/hexane, 4.3% methyl-t-butyl ether (MTBE)/hexane, or 2% acetonitrile/hexane. While Table 2.2 is primarily designed for unmodified adsorbents, it also can be used approximately for normal bonded-phase columns.

Selectivity Effects

Once the solvent strengths of the various binary mobile-phase mixtures (Table 4.8) have been determined for a particular sample, the effect of different

TABLE 4.8 Solvent Strength $\epsilon°$ vs. Mobile-Phase Composition for Cyano and Amino Columns

| | Percent by Volume (%v) Mixture for Indicated $\epsilon°$-Value[a] | | | | | |
| | Cyano Columns | | | Amino Columns | | |
$\epsilon°$	MC/Hex	ACN/Hex	MTBE/Hex	MC/Hex	ACN/Hex	MTBE/Hex
0.00	0	0	0	0	0	0
0.02	17	6	12	5	2	13
0.04	42	17	29	11	4	28
0.06	100	27	48	19	6	45
0.08		48	73	30	9	59
0.10		84	100	47	13	100
0.12				75	18	
0.14				100	25	

[a]MC, methylene chloride; Hex, Hexane; ACN, acetonitrile; MTBE, methyl-t-butyl ether.
Source: Reprinted with permission from Ref. (27).

mobile phases on band spacing can be determined in a manner similar to that for reversed-phase systems (Sect. 4.4). "Easy" and "average" samples are handled in a manner parallel to that for reversed-phase chromatography. As an example of "hard" samples, consider the separation of thirteen substituted naphthalenes on a silica column, using the scheme outlined above and summarized in Fig. 4.16. In this case, methylene chloride is the nonlocalizing solvent, methyl-t-butyl ether (MTBE) is the basic localizing solvent, and acetonitrile (with methylene chloride co-solvent, if needed for hexane solubility) is the nonbasic localizing solvent. Fig. 4.17 shows the separations obtained

TABLE 4.9 Mobile Phases Illustrating Solvent-Selectivity Effects in Normal-Phase HPLC for Runs of Figure 4.17

| | Solvent (%v) | | | |
Run No.	Hexane	MC	MTBE	ACN
1	42.2	57.8	0	0
2	95.8	0	4.2	0
3	87.0	10.0[a]	0	3.0
4	76.8	22.0	1.2	0
5	92.0	4.8	1.6	1.6
6	68.6	30.0	0	1.4
7	88.7	9.0	1.1	1.2

[a]Needed as a co-solvent for Hexane/ACN. Symbols as in Table 4.8.
Source: Reprinted with permission from Ref. (28).

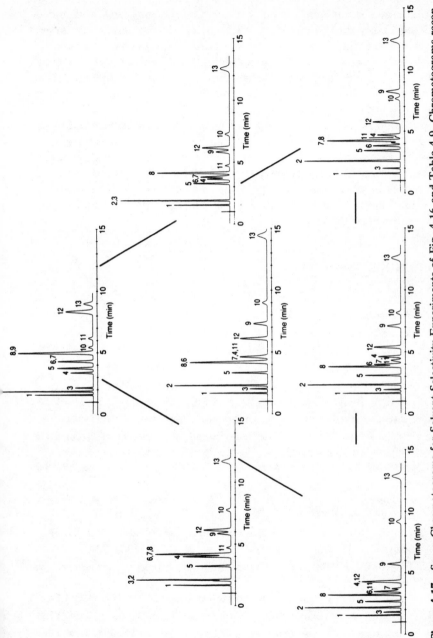

Figure 4.17 Seven Chromatograms for Solvent-Selectivity Experiments of Fig. 4.16 and Table 4.9. Chromatograms reconstructed from data in Ref. (28). Column 15 × 0.46-cm Zorbax-SIL, flow rate 2.0 mL/min; 35° C; Mobile phases as listed in Table 4.9; compounds are substituted naphthalenes: (1) 2-OCH$_3$, (2) 1-NO$_2$, (3) 1,2-(OCH$_3$), (4) 1,5-(NO$_2$), (5) 1-CHO, (6) 2-CO$_2$CH$_3$, (7) 1-CO$_2$CH$_3$, (8) 2-CHO, (9) 1-CH$_2$CN, (10) 1-OH, (11) 1-COCH$_3$, (12) 2-COCH$_3$, (13) 2-OH.

using the three different "binary" solvent mixtures (runs 1–3 in Table 4.9). Note that run 3 is not strictly binary, since a co-solvent is involved, but is considered binary for the purposes of selectivity. It is apparent that major changes in selectivity are occurring for the various mobile-phase mixtures. If one of these separations was adequate, search for an optimum solvent mixture would be concluded, yielding the optimized separation of Fig. 2.17b. Further improvement in the separation might be obtained by varying flow rate and column configuration (Sect. 2.5).

In the example of Fig. 4.17, the next step is to continue solvent optimization by running experiments 4–7. Further band-spacing changes can be achieved by examining the seven resulting chromatograms and making minor adjustments in concentrations to separate specific bands of interest. Alternatively, more exact predictions can be made using the computer methods described in Chapter 8. For an example of this general approach with a cyanophase column, see Ref. (29).

REFERENCES

1. C. F. Poole and S. A. Schuette, *Contemporary Practice of Chromatography*, Elsevier, Amsterdam, 1984, p. 375.
2. Sj. van der Wal and L. R. Snyder, *J. Chromatogr.*, *255* (1983) 463.
3. J. E. Eble, R. L. Grob, and L. R. Snyder, *J. Chromatogr.*, *384* (1987) 45.
4. V. V. Berry, *LC Mag.*, *2* (1984) 100.
5. Sj. van der Wal and L. R. Snyder, *Clin. Chem.*, *27* (1981) 1233.
6. F. Erni, H. P. Keller, C. Morin, and M. Schmitt, *J. Chromatogr.*, *204* (1981) 65.
7. *Solvent Guide*, Burdick & Jackson Laboratories, Muskeon, Mich., 1984.
8. Application sheets, E. I. duPont de Nemours and Co., Instruments Products Division, Wilmington, Del.
9. M. A. Quarry, E. I. duPont de Nemours and Co., Wilmington, Del, unpublished data, 1987.
10. M. A. Quarry, R. L. Grob, L. R. Snyder, J. W. Dolan, and M. P. Rigney, *J. Chromatogr.*, *384* (1984) 163.
11. L. R. Snyder, M. A. Quarry, and J. L. Glajch, *J. Chromatogr.*, *24* (1987) 33.
12. L. R. Snyder, J. W. Dolan, and M. P. Rigney, *LC-GC*, *4* (1986) 921.
13. J. L. Glajch, J. J. Kirkland, K. M. Squire, and J. M. Minor, *J. Chromatogr.*, *199* (1980) 57.
14. S. Sekulic, P. Haddad, and C. J. Lamberton, *J. Chromatogr.*, *363* (1986) 125.
15. R. W. Stout, J. J. DeStefano and L. R. Snyder, *J. Chromatogr.*, *261* (1983) 189.
16. M. T. W. Hearn, ed., *Ion-Pair Chromatography*, Marcel Dekker, New York, 1985.

17. J. H. Knox and R. A. Hartwick, *J. Chromatogr.*, *204* (1981) 3.

18. A. P. Goldberg, E. Nowakowska, P. E. Antle, and L. R. Snyder, *J. Chromatogr.*, *316* (1984) 241.

19. J. E. Eble, Ph.D. Thesis, University of Villanova, Villanova, Penn., 1987.

20. L. R. Snyder and J. J. Kirkland, *Introduction to Modern Liquid Chromatography*, 2nd ed., Wiley-Interscience, New York, 1979, Chapt. 9.

21. E. L. Wieser, A. W. Salotto, S. M. Flach, and L. R. Snyder, *J. Chromatogr.*, *303* (1984) 1.

22. L. R. Snyder and T. C. Schunk, *Anal. Chem.*, *54* (1982) 1764.

23. L. R. Snyder and J. J. Kirkland, *An Introduction to Modern Liquid Chromatography*, 2nd ed., Wiley-Interscience, New York, 1979, Chapt. 15.

24. J. L. Glajch, J. J. Kirkland, and W. G. Schindel, *Anal. Chem*, *54* (1982) 1276.

25. L. R. Snyder, J. L. Glajch, and J. J. Kirkland, *J. Chromatogr.*, *218* (1981) 299.

26. L. R. Snyder, in *High-Performance Liquid Chromatography. Advances and Perspectives Vol. 3*, Cs. Horvath, ed., Academic Press, New York, 1983, p. 157.

27. L. R. Snyder, *LC-GC*, *1* (1983) 478.

28. J. J. Kirkland, J. L. Glajch, and L. R. Snyder, *J. Chromatogr.*, *238* (1982) 269.

29. M. DeSmet, G. Hoogewijs, M. Puttemans, and D. L. Massart, *Anal. Chem*, *56* (1984) 2662.

5

DIFFICULT SEPARATIONS: THE USE OF OTHER SEPARATION VARIABLES AND PROCEDURES

Most samples can be successfully separated by the approach presented in Chapter 4. That is, once the HPLC method is selected (reversed-phase, ion-pair, or normal-phase chromatography), and the mobile-phase composition and column conditions are optimized, adequate resolution of the sample is usually achieved. However, the separation may be less than satisfactory in other respects. For example, the run time may be too long, and there may be many samples to analyze. In this case, further fine-tuning of the separation is worthwhile in order to reduce run time. Or, for some samples, the choice of mobile phases may be limited because of interference with detection below 210 nm. In such cases, method development may be restricted to reversed-phase HPLC with acetonitrile/water as the mobile phase. Under these conditions, failure to develop an adequate separation becomes more likely. Finally, some samples resist easy separation. The sample may contain a dozen or more components, or some of the compounds may be so chemically similar as to yield identical retention times in most HPLC systems.

Whenever mobile-phase optimization as in Chapter 4 is unsuccessful in adequately resolving a sample, additional separation variables must be explored. Column conditions can always be varied over wide limits to further improve the separation (i.e., optimizing N: see Sect. 2.5). However, difficult separations will normally require changes in band spacing (α) for further improvement. Table 5.1 summarizes some additional separation conditions that can be used for this purpose. The most attractive approach often is to select another HPLC method and start over, as discussed in Chapter 4. For example, if initial method development used reversed-phase separation, and the sample is composed of acids and/or bases, then ion-pair HPLC might be a good alternative. Now, the mobile-phase composition can be optimized by varying pH and the concentration of the ion-pair agent (see Sect. 4.4). Alternatively, such variables as column type, temperature, etc., can be used to alter band spacing.

TABLE 5.1   Additional Separation Variables to Change Band Spacing in HPLC

- Different HPLC method
- Ionic strength (buffer concentration)[a]
- pH[a] plus solvent optimization
- Column type (C-18, cyano, phenyl, diol, etc.)
- Same column type from different source[a]
- Mobile phase additives (amines and complexing agents)[a]
- Temperature[a]
- Additional organic solvents

[a]Mainly for ionizable samples (acids and bases).

Altering selectivity in these ways should rarely fail to achieve total resolution of the sample, except for quite complex mixtures (20 or more compounds). Samples of biological origin often fall into this category. When there is a large number of components in one sample, it is often difficult to find space in the chromatogram for all the individual (resolved) bands, even when the position of these bands can be varied over wide limits. As a general rule, there must be enough room in the chromatogram (peak capacity) for at least three times as many bands as are present in the sample. Some samples simply require more resolving power than a single HPLC separation can provide.

Complex samples are often best handled by some form of multidimensional chromatography, where an initial HPLC (or other) separation provides simpler fractions that can be adequately resolved in subsequent HPLC runs. When two or more sequential separations are used, band spacing can be independently adjusted in each HPLC run ("selectivity switching"). This approach is conveniently carried out using some form of column switching. We will briefly discuss multidimensional separations in this chapter.

While the total analysis of a complex sample is sometimes a separation goal, separation and analysis are often required for only one or a few components present in the sample. The approach in this case is similar to that for total sample analysis, but the goal is simpler to meet.

5.1 USE OF ANOTHER HPLC METHOD

In most cases, method development is started with reversed-phase HPLC. If separation is unsuccessful following mobile-phase optimization (see Sect. 4.3), then either ion-pair or normal-phase HPLC is usually an appropriate alternative. Table 1.3 can be consulted for general advice on how to proceed with developing a method using these primary HPLC methods. If the separation is still unsuccessful at this point, the secondary methods of Table 1.4 should also be considered. Usually, a change in the HPLC method (followed by appropriate solvent optimization) is the most powerful means available for changing band spacing.

If the problem is one or more unresolved band-pairs, the sample should be considered for any special characteristics that require a more specialized approach: high-molecular-weight components, mixtures of optical isomers, etc. Chapter 7 can be consulted for samples that fall into this category.

5.2 REVERSED-PHASE HPLC

If an inadequate separation is obtained after optimizing the mobile phase (see Chapt. 4), the next step is to examine other separation variables (see Table

5.1). At this point, a partial separation of the sample may have been attained, together with some understanding of how the separation varies with changes in mobile-phase composition. The first question that arises is: how can this information on mobile-phase selectivity be fruitfully combined with studies involving some new separation variable?

As a general rule, it is not profitable to optimize different separation conditions independently of each other. For example, assume that 46% acetonitrile/water gave the best separation of the sample (of all mobile phases studied), and that a change in temperature is considered as a possible way to improve resolution. Once a new temperature is selected, the best mobile-phase composition will likely change. This means that the mobile phase must be reoptimized when temperature (or any other separation variable) is changed. One exception is the situation where two compounds overlap for every mobile phase tried. In this case, some other separation parameter (column type, temperature, etc.) must be changed to resolve the difficult band pair. Mobile-phase composition need not be varied simultaneously, because in this instance, the mobile phase does not affect the separation of the critical band pair. Once the overlapping band-pair is adequately resolved, however, other bands may now be poorly separated. The separation must then be reoptimized with respect to all the important variables.

Ionic Strength and pH

For samples that contain acidic or basic compounds, retention can vary with both ionic strength and pH. An increase in ionic strength reduces the retention of protonated bases on column packings with ionized, residual silanols; sufficiently high salt concentrations can increase the retention of neutral compounds. A change in pH that increases the ionization of a sample compound will also usually reduce retention in a reversed-phase separation. These trends are illustrated by the following examples.

Figure 5.1 shows the usual effect of mobile-phase pH on the retention of acidic, neutral and basic molecules. Salicylic acid (compound 1 in Fig. 5.1) is a moderately strong acid, with a pK_a-value of 3. Therefore, at pH = 3, salicylic acid will be half ionized. At a pH two units lower (pH = 1), this compound will be completely protonated (nonionized), and its retention should be roughly twice as great, assuming that the ionized molecule is retained much less than is the nonionized compound. At a pH two units higher (pH = 5), the molecule will be almost completely ionized, and its retention should be at a minimum. Further increase in pH will have minimal effect on retention, because ionization of the molecule does not change. These changes in retention with pH for salicylic acid are confirmed by the data in Fig. 5.1.

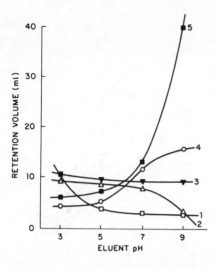

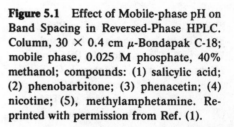

Figure 5.1 Effect of Mobile-phase pH on Band Spacing in Reversed-Phase HPLC. Column, 30 × 0.4 cm μ-Bondapak C-18; mobile phase, 0.025 M phosphate, 40% methanol; compounds: (1) salicylic acid; (2) phenobarbitone; (3) phenacetin; (4) nicotine; (5), methylamphetamine. Reprinted with permission from Ref. (1).

In similar fashion, retention of the weak acid phenobarbitone (compound 2, Fig. 5.1) shows little change until the mobile phase has a pH greater than 7. At this point, the molecule begins to ionize and its retention drops steeply. The retention of neutral molecules such as phenacetin (compound 3, Fig. 5.1) shows little dependence on pH. A strong base such as methylamphetamine (compound 5, Fig. 5.1, $pK_a = 8$) shows increasing retention above pH 8, as the molecule becomes progressively less ionized; retention decreases steeply at lower pH, but approaches a constant retention for pH < 6. The weak base, nicotine (compound 4, Fig. 5.1), shows retention increasing over the range 5 < pH < 9, suggesting a pK_a-value of about 7 in this system.

The retention dependencies of Fig. 5.1 show that pH can be a powerful factor in changing band spacing. In this particular example, all five bands are well separated at a pH of either 5 or 8; pH 5 is preferred because of shorter run time and increased column life. If the pK_a-values are known for a group of compounds to be separated by reversed-phase HPLC, and if these pK_a-values are different, then it is likely that a pH near (± 2 units) the average pK_a-value of the mixture should provide good separation. The reason is that pronounced changes in band spacing can be expected as pH is varied in the region of pH = pKa.

As an example, consider the bile acids whose reversed-phase retention is plotted vs. pH in Fig. 5.2. These compounds all have pK_a-values in the 5–6 range. It would be predicted, therefore, that a mixture of these compounds will be best separated at a pH not far from 5–6; e.g., 4 < pH < 7. In Fig. 5.2, five band-spacing changes are observed for this mixture as mobile-phase pH

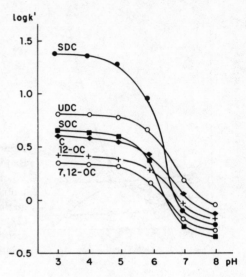

Figure 5.2 Effect of Mobile-Phase pH on Band Spacing for the Reversed-Phase HPLC Separation of a Bile Acid Mixture. Column, Nucleosil C-18; mobile phase, 45% acetonitrile/phosphate buffer; compounds: see Ref. (2). Reprinted with permission from Ref. (2).

varies from 5 to 7. Outside this pH range there is no further change in band spacing, as expected from the pK_a values of these solutes. The best separation occurs for a pH near 6.2, where band spacing is good, and the retention range is suitable ($1.5 < k' < 10$). Alternatively, pH = 4 also would afford good band spacing; the concentration of organic solvent could be adjusted to provide the desired k' range.

Figure 5.3 shows the effect of salt concentration (ionic strength) on the separation of a mixture of PTH-amino acids. In this mixture, the basic amino acids (lysine, arginine [R], histidine [H]) are protonated at the pH of the mobile phase, and probably interact with the column-packing silanols by ion exchange. Increasing the salt concentration decreases ion-exchange retention and decreases the retention of these basic compounds, resulting in a pronounced change in band spacing (Fig. 5.3a vs. 5.3b). With a complex mixture, simultaneously achieving a good band spacing for all bands can be a real challenge. By using a variable that affects only certain compounds in the mixture (salt concentration in Fig. 5.3), the overall optimization of retention is considerably simplified, much like the example of Fig. 2.19. Thus, to develop an optimized separation in Fig. 5.3, the three basic PTH-amino acids were not included in the initial retention optimization for this mixture. With these three compounds removed from the sample, retention optimization for the remaining 17 compounds was correspondingly easier. Once an acceptable band spacing for these 17 compounds was attained (by solvent and column optimization), the three basic PTH-amino acids could be shifted into

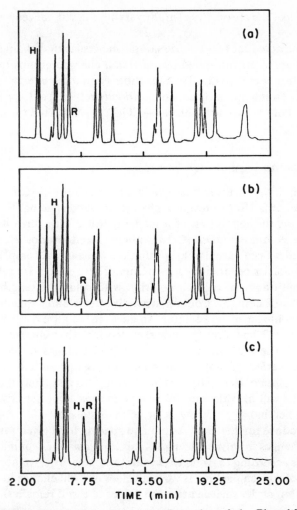

Figure 5.3 Effect of Ionic Strength on the Retention of the Phenylthiohydantoin (PTH) derivatives of Histidine (H) and Arginine (R). Column, 8 × 0.62-cm, 3-μm Zorbax Phenyl; Mobile-phase gradient: methanol 5.3%–13.2%, tetrahydrofuran 9.1%–25.8%, phosphate buffer (pH 3.2) 85.6%–61% in 15 min, then held at final concentrations for 15 min. (a) 12 mM buffer; (b) 8 mM buffer; (c) 4 mM buffer. Reprinted with permission from Ref. (4).

"empty" parts of the chromatogram, by varying the salt concentration in the mobile phase.

It should be noted for amphoteric sample molecules that changes in retention with pH are often controlled by *individual* charged groups in the molecule, not by the *net* charge (3). Thus, peptides containing one carboxyl group (acid) and one amino group (base) are *minimally* retained at intermediate pH values, where both groups are ionized ($-COO^-$, $-NH_3^+$) but the net molecular charge is zero.

Column-Packing Type

Many workers have observed that changing the column-packing type (e.g., phenyl instead of C-18) can result in changes in band spacing. For example, Fig. 2.20 shows the separation of a model mixture with three different reversed-phase columns: ODS (C-18), phenyl, and cyano. Here, solvent strength has also been varied to keep run time constant. A completely different separation order occurs for each column. These effects are related to (a) column strength and (b) column polarity, with column strength being the more important factor (5). Column strength affects band spacing, because more hydrophobic (stronger) columns (e.g., C-18) require a stronger mobile phase. Section 4.2 has already addressed the fact that changes in mobile-phase strength (percent organic) often result in changes in band spacing. However, if the effect of mobile-phase percent organic on band spacing has been studied, and no useful changes in α found, it is less likely that a change in column type will provide a useful change in separation selectivity.

On the other hand, differences in column polarity may still be large enough to produce further changes in band spacing for a given sample, independent of changes in solvent strength. Table 5.2 classifies common column-packing types according to strength (J value) and polarity (P value). When the surface area of a given packing is varied, values of J will change in proportion to the logarithm of the surface area, but values of P will remain the same. As

TABLE 5.2 Strength and Polarity of Common Reversed-Phase Packings

Column-Packing Type	Column Strength J		Column Polarity P
C-18	0.26	(strong)	−0.55
C-8	0.00		0.00
Phenyl	−0.54		1.24
TMS	−1.08		0.96
Cyano	−1.64	(weak)	1.00

Source: Reprinted with permission from Ref. (5).

column polarity P increases (different bonded phase), more polar compounds are retained preferentially compared with less polar compounds.

When using a new column-packing type, mobile-phase optimization must be repeated. It cannot be assumed that the same mobile phase that was optimum for the first column will also be optimum for the second column. This combined use of column type and mobile-phase optimization has allowed the development of a rapid isocratic separation of all 20 PTH-amino acids (4). The data in Table 5.2 show that column strength and polarity are quite different for C-18, phenyl (or benzyl), and cyano columns. Therefore, the results of mobile-phase optimization with each of these three column types should improve the chances of getting a good separation, compared with using any single column. Figure 5.4a outlines this approach to method development. Figures 5.4b–d show the separation of the PTH-amino acids with each of these three columns, using the same binary-solvent mobile phase in each case (methanol/water). Significant band-spacing changes occur for each column type.

Column-Packing Source

Differences in the types and concentrations of residual silanols exist for columns prepared in different ways and sold by different companies (see Sect. 3.1). Consequently, C-8 or C-18 columns from different sources often show marked changes in band spacing for a given sample. These retention differences occur mainly for polar compounds, particularly bases or cationic species. Figure 2.21 is a typical example of the kind of band-spacing variation that is possible with columns of the "same" type of packing obtained from different suppliers. Another example of a change in band spacing from one C-18 column to another is shown in Fig. 5.5. Here, a mixture of plant hormones was separated by reversed-phase gradient elution on two different columns, Hypersil ODS and Spherisorb ODS. The elution order for these five compounds is quite different in each case:

Hypersil (Fig. 5.5a)   ZOG < ZR < Z < IAA < ABA

Spherisorb (Fig. 5.5b)   IAA < ZR < ZOG < ABA < Z

Many workers routinely rely on C-8 or C-18 columns from different companies to provided needed band-spacing changes in reversed-phase separations. However, with this strategy, it is usually found that column-to-column variability is a severe problem. The reason is that no column supplier is currently able to precisely control column reproducibility, especially the silica support and the influence of silanols on retention. Until this situation changes, *we*

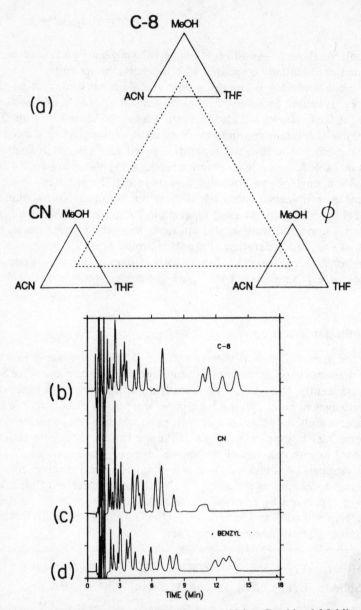

Figure 5.4 Approach to Method Development Using Combined Mobile- and Stationary-Phase Optimization. (a) Schematic diagram of experimental approach using a C-8, CN, and phenyl (ϕ) column with MeOH, ACN, and THF in water. Chromatograms (b,c,d) are for mixtures of 20 PTH-amino acids; mobile phase-pH 2.1 phosphoric acid, methanol, acetonitrile, tetrahydrofuran (66/13.3/10.7/10); flow rate, 2.0 mL/min, temperature, 35° C, columns: (b) 8 × 0.62 cm Zorbax C-8, (c) 8 × 0.62 cm Zorbax CN, (d) 8 × 0.62 cm Zorbax Phenyl. Reprinted with permission from Ref. (4).

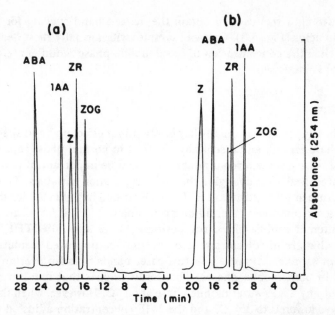

Figure 5.5 Separation of Plant Hormones by Reversed-Phase Gradient Elution. Effect of column type on band spacing. Mobile phase, gradient from 0 to 50% methanol/water (pH 3.3). Compounds: indole-3-acetic acid, IAA; zeatin, Z; zeatin riboside, ZR; abscisic acid, ABA; zeatin-o-glucoside, ZOG. (a) Hypersil ODS column; (b) Spherisorb ODS column. Reprinted with permission from Ref. (6).

recommend that silanol effects be avoided so far as possible by the use of suitable mobile-phase additives (see Sect. 3.3). While this approach eliminates the possible use of silanols to control band spacing, other alternatives for controlling α are preferred for long-term applications (see Chapter 4 and following discussion).

Table 3.1 characterizes different commercial packings in terms of their silanol acidity. These data can be useful in selecting columns with different selectivity. Column packings also vary in the relative coverage (percent carbon) of the silica surface by the bonded phase. For some samples, column selectivity may correlate with the concentration of bonded organic ligand on the silica surface (7,8).

Mobile-Phase Additives

Compounds that interact with the silanols of the column packing can be added to the mobile phase to control band spacing, in much the same way that ionic strength was used to obtain the desired band spacing for the 20 PTH-amino acids (Fig. 5.3). Cationic samples will normally show decreased retention when the concentration of basic mobile-phase modifiers (e.g., triethylamine) is increased.

Temperature

A change in temperature normally has only a minor effect on band spacing in reversed-phase HPLC, and essentially no effect in normal-phase separations (see Sect. 2.4). However, even a minor effect can be important, if two bands are still unresolved after changing the major separation variables. The effect of temperature in reversed-phase HPLC is illustrated in Table 5.3 for the separation of a six-component steroid mixture. Values of k' at 60° C are shown for two different mobile-phase compositions, 15% and 30% THF/water. Normally, changes in solvent strength can provide improved resolution for less-difficult samples. However, in this case, bands 5 and 6 overlap (same k'-values) in both mobile phases. This result implies that band overlap is expected for any THF/water mobile phase at 60° C. However, when the temperature was lowered to 35° C, and the THF concentration adjusted to give the same run time as for the 15%-THF mobile phase at 60° C, bands 5 and 6 were separated (with an α-value of 1.08). This sample cannot be resolved by changing solvent strength, but it can be resolved by varying separation temperature.

As was pointed out in Sect. 2.4, a change in temperature can lead to even more significant changes in band spacing, whenever the molecular weight or

TABLE 5.3   Effect of Temperature on Relative Retention of a Six-Component Steroid Sample[a]

Compound	k' (60° C) 15% THF	k' (60° C) 30% THF	k' (35° C), 20% THF	α^b
1. Prednisone	4.87	0.75	5.13	1.14
2. Cortisone	5.59	0.93	5.86	1.27
3. Hydrocortisone	7.37	1.22	7.43	1.49
4. Dexamethasone	11.55	1.91	11.04	1.28
5. Corticosterone	15.45	2.60	14.16	1.08
6. Cortexolone	15.45	2.60	15.28	—

[a]8 × 0.62-cm, Zorbax Golden Series C-8; 3-μm particles; 3 mL/min.
[b]Values of α for 35° C separation.
Source: Reprinted with permission from Ref. (9).

shape of two compounds differs significantly. However, compounds whose structures differ this much usually can be separated by changes in other experimental conditions.

Other Organic Solvents

The previous discussion of the solvent-selectivity triangle (Sect. 2.3) suggests that methanol, acetonitrile, and THF are preferred solvents for controlling band spacing in reversed-phase HPLC. Other solvents in the same regions of the selectivity triangle (e.g., dioxane and THF) are expected to yield similar α-values, and generally are not useful alternatives. Solvents from other parts of the selectivity triangle are often not miscible with water, and water-immiscible solvents are rarely used in reversed-phase HPLC. For these and other reasons, most workers have chosen methanol, acetonitrile, and THF (with water) as useful and practical solvents for controlling band spacing in reversed-phase HPLC.

If additional solvents are considered as co-solvents to be mixed with some starting organic/water mixture, the choice of organic solvents is considerably expanded. This approach allows the use of solvents near the corners of the selectivity triangle (e.g., ethers, chloroform, and methylene chloride), which considerably increases the possible range of acidity, basicity, and dipolarity available for the final mobile phase. This technique also increases the possibility of a better band spacing for a given sample. A few studies of this type have been reported (e.g., Ref. 10 and other references cited there), but with somewhat ambiguous conclusions. Figure 5.6 shows an example of an improved separation (Fig. 5.6b) that resulted from the addition of 10% ethyl ether to 35% acetonitrile/water as the mobile phase (Fig. 5.6a). The inconvenience of using water-immiscible solvents for this purpose is another factor to consider. However, for difficult separations, where the standard three organic solvents (ACN, MeOH, and THF) have not provided adequate resolution, use of additional solvents (even those that are not water-miscible) may be a worthwhile alternative.

However, the selectivity effects obtained with other organic modifiers in reversed-phase separations can be deceiving. For example, the same study (10) that provided the example of Fig. 5.6, furnished additional information on the use of ethers as mobile-phase solvents for other steroid samples. These data are summarized in Table 5.4, where k'-values are given for each compound and mobile phase. In each case, a particular ether was added to the original methanol/water mobile phase. At first glance, it appears that each of these six ethers provides uniquely different selectivity! This effect is seen more clearly by simply listing the separation sequence for each ether:

| | Separation Sequence[a] | | | | | | | k′ (Bands 4,5) |
Solvent	(1 = first band; 7 = last band)							
No ether	1	2	3 =	4 =	5	6	7	9.3
Dioxane	1	2 =	3 =	4 =	5	6	7	3.8
Dimethoxyethane	2	1	3 =	4 =	5	6	7	3.8
Cyclohexene oxide	2	1	6	5	3 =	4	7	—
THF	2	6	1	5	3 =	4	7	2.7
Ethyl ether	2	6	1	5	3 =	4	7	2.1
Isopropyl ether	2	6	7	5	1	4	3	1.8

[a]See Table 5.4; 3 = 4 means that for this band-pair, $\alpha < 1.04$ (difficult or no separation).

However, the reversed-phase solvent strength of these ethers increases from dioxane to isopropyl ether, as seen by the average k′-values for bands 4 and 5 in Table 5.4. Therefore, a solvent-strength band-spacing effect is being superimposed onto the ether-related selectivity of the mobile phase. In fact, a regular change in separation sequence occurs from the first mobile phase (no ether) to the last (isopropyl ether). Thus, bands 1, 3, and 4 show increasing retention relative to other steroids, while bands 2, 6, and 7 show decreasing retention. These changes in band position correlate with solvent strength (average k′). This effect suggests that any of these stronger ethers (THF, ethyl

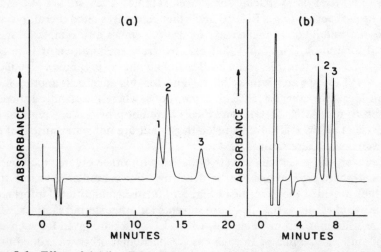

Figure 5.6 Effect of Adding Ethyl Ether to an Acetonitrile/Water Mobile Phase in the Separation of Estrogens. (a) 25-cm, 10-μm C-8 column; 35%v acetonitrile/water; 2 mL/min; (b), same, 10%v ethyl ether added to mobile phase of (a). Reprinted with permission from Ref. (10).

TABLE 5.4 Changes in Band Spacing on Adding Different Ether Solvents to a Methanol/Water Mobile Phase[a]

	k'-Values for Different Ethers Added[b]						
Solute	A	B	C	D	E	F	G
1. Estrone	7.88	3.53	3.49	5.27	2.32	1.94	1.79
2. Norethindrone	8.65	3.71	3.32	4.59	1.92	1.47	0.52
3. Estradiol-17α	9.27	3.91	3.91	6.35	2.82	2.19	1.87
4. 17β-Ethinyl estradiol	9.29	3.82	3.88	6.5	2.82	2.18	1.95
5. Estradiol-17α	9.35	3.77	3.79	6.06	2.65	2.06	1.59
6. Testosterone	11.3	4.53	4.12	5.47	2.09	1.59	0.56
7. Progesterone	23.9	9.0	7.82	10.9	4.0	2.97	1.42

[a]Steroid sample, C-8 column.
[b]Ether solvents: A, no added solvent (45% methanol/water); B, dioxane; C, 1,2-dimethoxyethane; D, cyclohexene oxide (2%); E, THF; F, ethyl ether; G, isopropyl ether. Unless noted otherwise, 10% of ether solvent was added to the mobile phase.
Source: Reprinted with permission from Ref. (10).

ether, and isopropyl ether) could provide similar separations, providing that the percent ether in each mobile phase was adjusted to give constant run times. Note also that the only ether to provide complete resolution of all bands (Table 5.4) is the strongest solvent: isopropyl ether. However, it seems likely that a higher concentration of ethyl ether would provide a similar separation. Consequently, there may not be any unique selectivity effects for the various ethers of Table 5.4.

Table 5.5 summarizes the different approaches to changing band spacing for reversed-phase HPLC. Separation variables are arranged in rough order of decreasing usefulness. Thus, changes in solvent strength (percent organic) and solvent type (methanol, acetonitrile, and THF), as discussed in Sect. 4.3, should be attempted first. Then, other variables can be explored in turn, until an acceptable separation is achieved.

5.3 ION-PAIR HPLC

Temperature

Temperature changes in ion-pair chromatography often lead to changes in band spacing (Sect. 2.4). The reason is that sample retention depends on the relative ionization of each solute, and this can change with temperature (independently of the retention of charged and uncharged molecules as a function of temperature). However, when a sample compound is ionized (or non-

TABLE 5.5   Controlling Band Spacing in Reversed-Phase HPLC

Variable[a]	Comment
Percent strong solvent	E.g., vary the percent acetonitrile/water; see Sect. 4.2 (first choice).
Solvent type	Methanol, acetonitrile, and THF; see Sect. 4.3.
Different HPLC method	Ion-pair or normal-phase HPLC; see Tables 1.3, 5.6, and 5.7.
Ionic strength	Use if sample contains basic compounds.
pH	Use if sample contains acids or bases.
Column type	C-8 (or C-18), cyano, and phenyl are good choices.
Column from different sources	C-8 (or C-18) columns that differ in carbon loading or acidity, see Table 3.1.
Amine additives	Vary triethylamine concentration.
Temperature	Varied between ambient and 60° C (last choice).

[a]Separation variables listed in approximately decreasing priority.

ionized) to the extent of 95% or more, a moderate change in temperature (e.g., by 20° C) should have only a small effect on relative ionization or retention. Changes in band spacing as a result of a change in temperature are most likely for compounds which have pK_a-values within ± 1 unit of the mobile-phase pH.

Figure 5.7 shows an example of changes in band spacing with temperature for a mixture of biogenic amines, separated by ion-pair HPLC. Figure 5.7a plots log k' vs. reciprocal absolute temperature ($1/T_K$) for these amines. The last two compounds (3-MT, 5-HT) reverse positions at an intermediate temperature. The corresponding chromatograms at 20° C and 45° C are shown in Fig. 5.7b, confirming this behavior. The relative spacing of other compounds in the sample also changes as temperature is varied (e.g., DA vs. HIAA), but less dramatically. For a similar example, see Ref. 12.

Mobile-Phase Additives

The separation of basic compounds by ion-pair HPLC appears to be less subject to silanol effects than is reversed-phase chromatography. Presumably, the sorption of the ion-pair agent onto the reversed-phase column (see Sect. 4.4) partly shields residual silanols and minimizes silanol–sample interactions. However, these silanol effects are not completely eliminated in ion-pair HPLC. When basic compounds are observed to tail in ion-pair systems (due

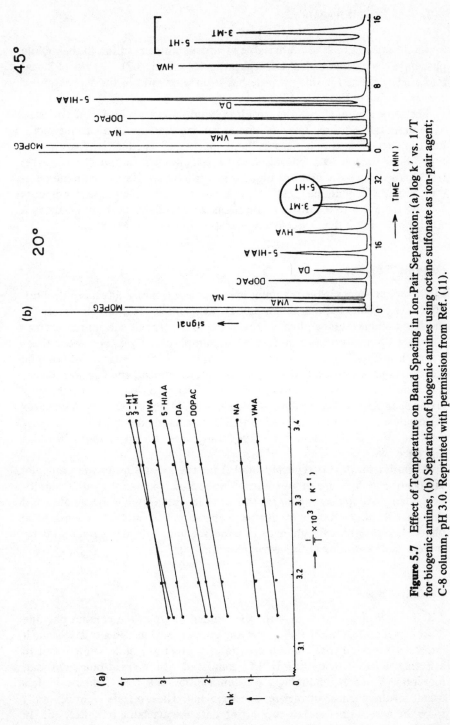

Figure 5.7 Effect of Temperature on Band Spacing in Ion-Pair Separation; (a) log k' vs. 1/T for biogenic amines, (b) Separation of biogenic amines using octane sulfonate as ion-pair agent; C-8 column, pH 3.0. Reprinted with permission from Ref. (11).

139

to silanol interactions), it is advisable to add an amine modifier to the mobile phase, just as for reversed-phase separations (see Sect. 3.2). The use of 20–30 mM triethylamine for this purpose has been successful in the separation of certain aniline derivatives (13).

The same silanol effects that result in band tailing can also be manifested as changes in retention. The addition of varying amounts of an amine modifier can then result in changes in band spacing. This effect is illustrated in Fig. 5.8 for the ion-pair separation of a six-component sample of amines related to the drug pafenolol (see Fig. 1.2 for structures). Here, the amine modifier, dimethyloctylamine (DMOA), is added to the mobile phase in amounts varying from 0.5 to 3 mM, and three changes in band position occur as a result of changing the DMOA concentration in the mobile phase.

Organic Solvent

Limited data suggest that changing the organic solvent usually has only a minor effect on band spacing in ion-pair HPLC. This effect is just the reverse of the situation in reversed-phase systems, where the organic solvent plays a major role in optimizing band spacing. One example of the organic solvent effect is shown in Fig. 5.9, for the ion-pair separation of this mixture of tricyclic antidepressants (amines). During method development, the organic solvent was chosen first, then pH and the concentration of ion-pair agent were optimized as in Sect. 4.4 (for each organic solvent). Minor changes in band spacing result for these three different organic solvents (methanol, Fig. 5.9a; acetonitrile, Fig. 5.9b; THF, Fig. 5.9c), but there are no changes in band sequence.

A similar result was reported in (14) for the separation of the pafenolol sample of Fig. 5.8; use of acetonitrile instead of 1-pentanol gave no significant changes in band spacing for this sample. Interestingly, it was observed that varying the solvent strength (concentration of pentanol in the mobile phase) led to significant changes in band spacing, with two compounds reversing their positions in the chromatogram.

Column Type

The ion-pair agent tends to cover the bonded phase of the column packing (Sect. 4.4), and to mask the underlying support surface. For this reason, it might be expected that column type (and source) will be less important in affecting retention in ion-pair HPLC, compared with reversed-phase chromatography. Similarly, changing the column in ion-pair HPLC should be less useful for the purpose of varying band spacing. These effects seem generally to be the case, although few supporting data are available (cf., Ref. 15). In

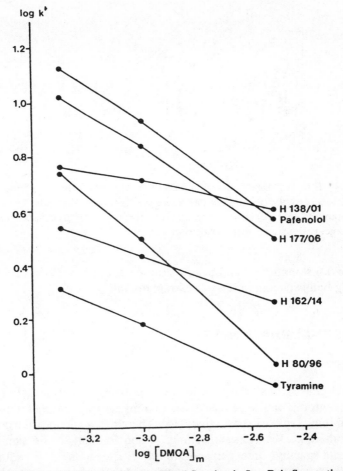

Figure 5.8 Effect of Amine Additive on Band Spacing in Ion-Pair Separation of Pa-fenolol Sample (Fig. 1.2). C-8 column; mobile phase: pentanol/water, pH = 2; di-methylcyclohexane sulfonate ion-pair agent. Reprinted with permission from Ref. (14).

any case, the use of column silanol effects to change band spacing is a less attractive alternative, just as for reversed-phase systems (Sect. 5.2).

Ionic Strength

Further changes in band spacing in ion-pair HPLC can be achieved by vary-ing mobile-phase ionic strength. However, these changes are similar to those that can be achieved by varying the ion-pair agent concentration (13). There-fore, ionic strength is less often used to optimize resolution in ion-pair HPLC.

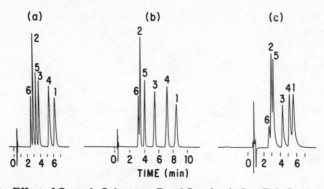

Figure 5.9 Effect of Organic Solvent on Band Spacing in Ion-Pair Separation of Tricyclic Antidepressants. pH and ion-pair agent concentration optimized for (a) methanol as organic solvent, (b) acetonitrile, and (c) THF; C-8 column, hexane sulfonate as ion-pair agent. Reprinted with permission from Ref. (13).

Table 5.6 summarizes and prioritizes the various alternatives for further changing band spacing in ion-pair chromatography.

5.4 NORMAL-PHASE HPLC

Column Type

In normal-phase HPLC, the polar interactions between sample molecules and functional groups on the surface of the column packing play a dominant role in determining separation selectivity (16). Therefore, a change from one polar bonded-phase packing to another (e.g., cyano vs. diol) should significantly affect band spacing. An example of this effect is seen in Fig. 5.10 for a normal-phase separation to assay a herbicide-metabolite residue in a wheat extract. Figure 5.10a shows the chromatogram obtained with a cyano column in which the fraction containing the herbicide metabolite was trapped (column switching) and diverted to a second column (diol phase), as shown in Fig. 5.10b. Several distinct bands were resolved on the diol column, whereas all of these compounds co-eluted from the cyano column. This same study also showed that a narrow fraction from the diol column (different sample, Fig. 5.10c) could be resolved into several bands on an unmodified silica column (Fig. 5.10d). In Figs. 5.10c,d different band spacings are found on these three normal-phase columns (cyano, diol, and silica) for extracts of green oats.

Other studies (e.g., Refs. 18,19) have also shown marked differences in the selectivity of cyano- and amino-phase and silica columns, as would be expected for the differing functionality of these phases (cf. solvent-selectivity triangle, Fig. 2.16).

TABLE 5.6   Controlling Band Spacing in Ion-Pair HPLC

Variable[a]	Comment
Percent strong solvent	E.g., vary percent methanol/water; see Sect. 4.2 (first choice); choose initial pH near average pK_a-value of sample; use intermediate concentration of ion-pair agent.
pH and concentration of ion-pair agent	$2.5 < pH < 7$; 0–20 mM TBA or 0–200 mM hexane sulfonate; see Sects. 4.4 and 9.5.
Temperature	Varied between ambient and 60° C.
Mobile-phase additives	For the separation of basic samples using hexane sulfonate, add TEA (up to $1/5$ the concentration of ion-pair agent).
Solvent type	Acetonitrile or THF to replace methanol (methanol is usually preferred for buffer solubility).[b]
Ionic strength	Vary buffer concentration or add salt to mobile phase.
Column type and source	Less promising.

[a]Separation variables are listed in approximately decreasing priority.
[b]Check for possible solubility problems when formulating ion-pair mobile phases that include higher concentrations of acetonitrile or THF.

Other Organic Solvents

Section 4.5 describes the use of three polar organic solvents to control band spacing in normal-phase HPLC: (a) a nonlocalizing solvent such as methylene chloride or chloroform, (b) a basic localizing solvent such as an ether, and (c) a nonbasic localizing solvent such as ethyl acetate or acetonitrile. These three solvents plus a nonpolar solvent such as hexane can achieve most of the solvent-related selectivity that is possible.

However, other organic solvents, especially alcohols, add additional possibilities for changing band spacing. Since it is easy to scout these options using TLC, this approach is an alternative for further adjusting band spacing (for trial-and-error use of different solvents in TLC, see Chapt. 9 of Ref. 20).

Temperature

Usually, a change in temperature will have, at most, a minor effect on band spacing in normal-phase separations (20); therefore, temperature change is rarely used for this purpose. An example of this effect is shown in Fig. 5.11a, for the separation of a mixture of organic compounds on silica. Band spacing

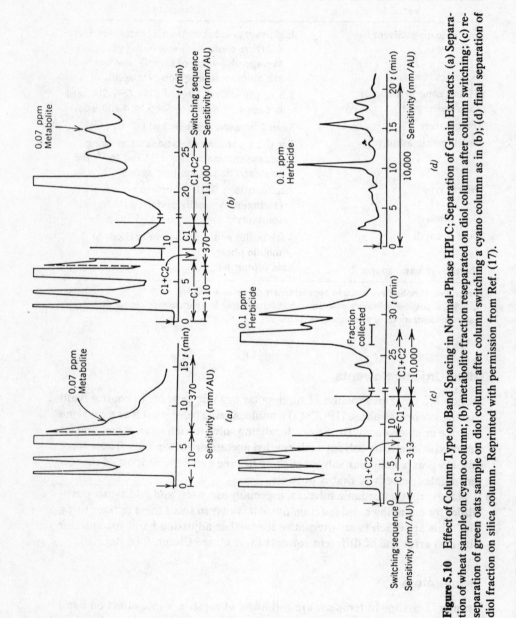

Figure 5.10 Effect of Column Type on Band Spacing in Normal-Phase HPLC; Separation of Grain Extracts. (a) Separation of wheat sample on cyano column; (b) metabolite fraction reseparated on diol column after column switching; (c) reseparation of green oats sample on diol column after column switching a cyano column as in (b); (d) final separation of diol fraction on silica column. Reprinted with permission from Ref. (17).

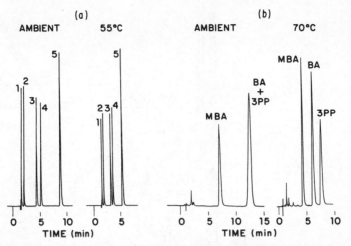

Figure 5.11 Effect of Temperature on Normal-Phase HPLC Column, 25 × 0.46-cm silica; compounds, 1, nitrobenzene; 2, methylbenzoate; 3, benzaldehyde; 4, acetophenone; 5, α-methylbenzyl alcohol (MBA); benzyl alcohol (BA); 3-phenyl-1-propanol (3PP). (a) methylene chloride mobile phase; (b) 2% acetonitrile/hexane mobile phase. Reprinted with permission from Ref. (20).

is essentially identical at 25° C and 55° C. Temperature-related changes in α are more often found for mobile phases that contain localizing solvents such as acetonitrile, as illustrated in Fig. 5.11b for a different sample; these effects are probably due to the desorption of polar solvents, at higher temperatures. However, it is likely that such band-spacing changes can be better approached by solvent optimization, as described in Section 4.5.

TABLE 5.7 Controlling Band Spacing in Normal-Phase HPLC

Variable[a]	Comment
Percent strong solvent	E.g., vary percent methylene chloride/hexane; see Sect. 4.2 (first choice).
Solvent type	Methylene chloride (chloroform), MTBE (THF), acetonitrile (ethyl acetate), methanol; see Sect. 4.5.
Column type	Cyano (first choice), diol, silica, amino.
Other solvents	For silica as packing, check other solvents using TLC and the solvent triangle (Fig. 2.16) as guide.
Temperature	Varied between ambient and 60° C.

[a]Separation variables are listed in approximately decreasing priority.

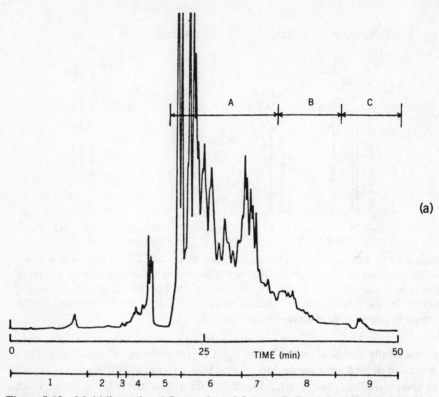

(a)

Figure 5.12 Multidimensional Separation of Solvent-Refined Coal Fraction by Reversed-Phase Gradient Elution. (a) Gradient separation on 25-cm, 5-μm C-18 column; (b) separation of fraction A from (a) on 1.8-m, 3-μm C-18 column. Reprinted with permission from Ref. (21).

Table 5.7 summarizes and prioritizes the various alternatives for further changing band spacing in normal-phase HPLC.

5.5 MULTIDIMENSIONAL TECHNIQUES

Multidimensional chromatography uses two (or more) different chromatographic separations to achieve the resolution of one or more compounds in the original sample (e.g., Fig. 5.12). Sample cleanup or pretreatment prior to HPLC separation is one form of multidimensional separation (Sect. 1.1), although this usually involves a low-resolution separation as the first step. Oc-

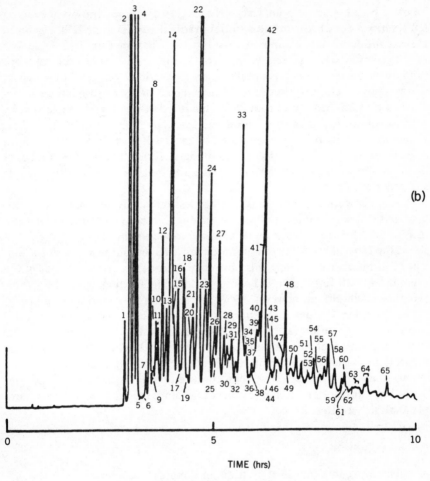

(b)

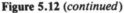

0 5 10

TIME (hrs)

Figure 5.12 (*continued*)

casionally, multidimensional separation is used for the total analysis of a very complex sample; for example, one containing several hundred components. More often this approach is used for the separation and analysis of one or more components in a complex sample matrix (e.g., trace analysis).

Total Sample Analysis

One technique for the complete analysis of complex samples is to use an initial gradient separation that elutes all of the peaks in a relatively short time, followed by a high-resolution (large N-value) isocratic separation of individual fractions obtained from this initial gradient run (similar HPLC column and conditions). An example of this approach is shown in Fig. 5.12. The starting sample was a solvent-refined coal, which probably has several thousand different components. A preliminary separation by normal-phase chromatography on alumina produced a fraction containing neutral polycyclic aromatics that was still quite complex. This fraction was further separated by reversed-phase step-gradient elution into fractions A–C shown in Fig. 5.12a. Fraction A was then further separated by reversed-phase HPLC on a 1.8-m, 3-μm column (N = 200,000) as shown in Fig. 5.12b. While a large number of bands resulted, the separation time was quite long (10 hr), and the sample still was not fully resolved.

A more powerful technique is to place less emphasis on large plate numbers, but instead combine *different* separation methods. An example is given in Fig. 5.13 for the separation of acidic compounds in urine. Figure 5.13a shows the separation of the initial sample by reversed-phase gradient elution. The total number of compounds obviously exceeds the peak capacity of the separation, with few individual compounds being adequately resolved. This sample was simplified, however, by an anion-exchange separation, as in Fig. 5.13b. Fraction B from this separation was further separated by reversed-phase gradient elution (as in Fig. 5.13a), as shown in Fig. 5.13c. Now, the overall resolution of this sample is much improved. Thus, the use of serial separations based on different separation principles (ion-exchange vs. reversed-phase) is more effective than the use of similar methods (as in Fig. 5.12). For a further discussion of multidimensional separation and total sample analysis, see Ref. 23.

Column Switching

A much more common use of multidimensional HPLC analysis is for the separation and determination of one or a few sample components (often present in low concentration). In these cases, it is necessary either to remove large-concentration interferences in the first separation, or to achieve good resolution of the analyte(s) in the final separation. A good example of this approach is seen in Figs. 5.10c,d, where successive separations are carried out on three different normal-phase columns. The herbicide-metabolite fraction from the initial cyano column is reseparated on a diol column, and then further separated on a silica column.

Fractions from an initial separation can be manually collected and reinjected for the second separation; however, this approach is tedious and often imprecise. A better alternative is the use of column switching, as in the examples of Fig. 5.10. For a general discussion of column-switching in HPLC, see

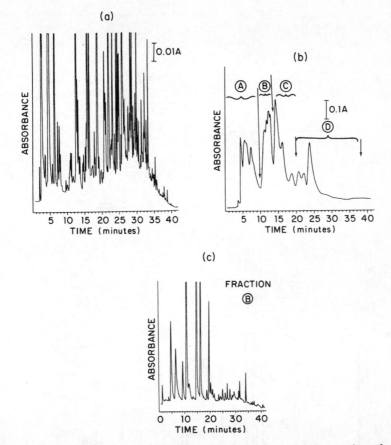

Figure 5.13 Multidimensional Separation of Urine Acids. (a) Separation of total sample by reversed-phase gradient elution; (b) separation of total sample by anion-exchange chromatography; (c) separation of fraction B from (b) by reversed-phase gradient elution. Reprinted with permission from Ref. (22).

Refs. (24–26). It should be noted that column switching can be carried out with different column packings (as in Fig. 5.10), or with different mobile phases. Figure 5.14 is an example of the determination of the chiral isomers of various amino acids in mixtures of these compounds. The HPLC system is shown in Fig. 5.14a, with resulting separations in Figs. 5.14b and c. Two independent HPLC systems are connected by a switching valve. In the first separation, reversed-phase gradient elution achieves the separation of individual amino acids, but without chiral separation. Individual amino-acid ra-

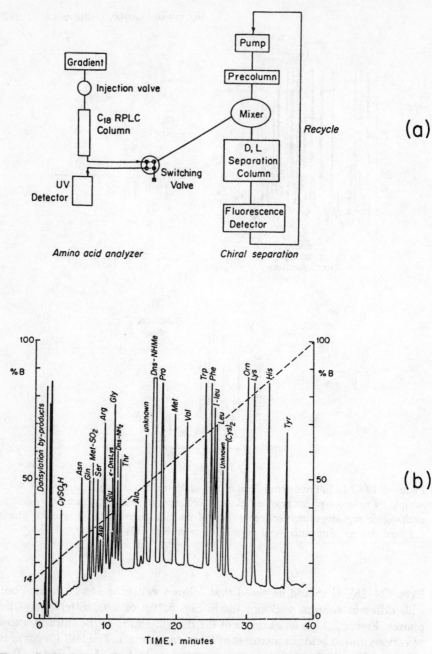

Figure 5.14 Column Switching for the Analysis of Individual D- or L-amino Acids. (a) Column-switching system; (b) separation of individual amino acids (as D/L mixtures) on first column; (c) separation of each amino acid into D- and L-isomers on second column. Reprinted with permission from Ref. (27).

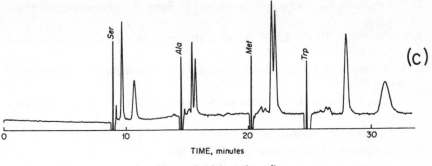

Figure 5.14 (*continued*)

cemates (D/L mixture) are transferred by the switching valve to the second system, where a chiral complexing agent has been added to the mobile phase (other conditions similar). The resulting separation of each amino acid into its D- and L-isomers is shown in Fig. 5.14c.

Column switching can also be used with different HPLC methods, for example, size-exclusion chromatography, followed by reversed-phase separation, as in Ref. (28). When all of the of options (column, mobile phase, and method) available for column switching are considered, it is obvious that many different variables can be used to achieve the resolution of a given compound from the rest of the sample mixture. Column switching as described so far is based mainly on differences in band spacing between the two (or more) separations. Any of the variables of Tables 5.5–5.7 can be changed between the two separations, allowing the separate optimization of band spacing in each separation. The preceding discussion of band-spacing control for a single separation applies to each of the separations used in column switching.

REFERENCES

1. P. J. Twitchett and A. C. Moffat, *J. Chromatogr., 111* (1975) 149.
2. D. S. Lu, J. Vialle, H. Tralongo, and R. Longeray, *J. Chromatogr., 268* (1983) 1.
3. M. T. W. Hearn, in *High-Performance Liquid Chromatography. Advances and Perspectives, Vol. 3*, Cs. Horvath, ed., Academic Press, New York, 1983.
4. J. L. Glajch, J. C. Gluckman, J. G. Charikofsky, J. M. Minor, and J. J. Kirkland, *J. Chromatogr., 318* (1985) 23.
5. P. E. Antle and L. R. Snyder, *LC Mag., 2* (1984) 840.
6. V. Sjut and M. V. Palmer, *J. Chromatogr., 270* (1983) 309.

7. B. Buszewski, K. Sebekova, P. Bozek, and D. Berek, *J. Chromatogr., 367* (1986) 171.

8. I. Wouters, S. Hendrickx, E. Roets, J. Hoogmartens and H. Vanderhaeghe, *J. Chromatogr., 291* (1984) 59.

9. M. A. Quarry, E. I. duPont de Nemours and Co., Wilmington, Del., unpublished data, 1987.

10. G. J.-L. Lee, R. M. K. Carlson, and S. Kushinsky, *J. Chromatogr., 212* (1981) 108.

11. N. Lammers, J. Zeeman, and G. J. deJong, *J. HRC CC, 4* (1981) 444.

12. J. H. Kennedy, *J. Chromatogr., 281* (1986) 288.

13. A. P. Goldberg, E. Nowakowska, P. E. Antle, and L. R. Snyder, *J. Chromatogr., 316* (1984) 241.

14. S. O. Jansson and S. Johansson, *J. Chromatogr., 242* (1982) 41.

15. C. Gonnet, C. Bory, and G. Lachatre, *Chromatographia, 16* (1982) 242.

16. L. R. Snyder, in *High-Performance Liquid Chromatography. Advances and Perspectives, Vol. 3*, Cs. Horvath, ed., Academic Press, 1983, p. 157.

17. J. F. K. Huber, I. Fogy, and C. Fioresi, *Chromatographia, 13* (1980) 408.

18. L. R. Snyder and T. C. Schunk, *Anal. Chem., 54* (1982) 1764.

19. A. W. Salotto, E. L. Weiser, and L. R. Snyder, Pace University, Pleasantville, NY, unpublished data, 1986.

20. L. R. Snyder and J. J. Kirkland, *Introduction to Modern Liquid Chromatography*, 2nd ed., Wiley-Interscience, New York, 1979, pp. 390–392.

21. M. Novotny, A. Hirose, and D. Wiesler, *Anal. Chem., 56* (1984) 1243.

22. E. L. Mattinz, J. W. Webb, and S. C. Gates, *J. Liq. Chromatogr., 5* (1982) 2343.

23. B. L. Karger, L. R. Snyder, and Cs. Horvath, *An Introduction to Separation Science*, Wiley-Interscience, New York, 1973, Chapt. 19.

24. L. R. Snyder and J. J. Kirkland, *Introduction to Modern Liquid Chromatography*, 2nd ed., Wiley-Interscience, New York, 1974, Chapt. 16.

25. R. E. Majors, *LC Mag., 2* (1984) 358.

26. K. Ramsteiner, *Int. J. Environ. Anal. Chem., 25* (1986) 49.

27. Y. Tapui, N. Miller, and B. L. Karger, *J. Chromatogr., 205* (1981) 325.

28. J. A. Apfel, T. V. Alfredson, and R. E. Majors, *J. Chromatogr., 206* (1981) 43.

6

GRADIENT ELUTION

In gradient elution the composition of the mobile phase changes during the separation. Usually a binary-solvent mobile phase is used, with the strong solvent B (e.g., methanol in reversed-phase HPLC) increasing in concentration during the gradient. We will refer to the concentration of this solvent as %-B in the following discussion.

Gradient elution has long been regarded as a research technique. Until recently, gradient elution was generally excluded from the routine laboratory, except for special applications with a wide retention range (isocratic k′ < 1 and k′ > 20 for different compounds in the same sample). An example of this is the analysis of amino acid mixtures. This situation developed as a result of several facts or perceptions:

1. Routine labs sometimes do not have HPLC equipment that is suitable for gradient separations.

2. Gradient-elution methods are believed to be less precise, compared with isocratic methods.

3. Gradient elution is a more complex technique than isocratic elution; therefore, method development is more difficult.

4. Gradient elution necessarily means long run times that are unsuitable for routine application, especially when considering the need for column reequilibration after each run.

5. Gradient-elution procedures are instrument-specific; methods developed on one gradient system may not perform the same way on another instrument model.

6. Quantitation is further complicated by the need for very pure mobile phases. Baselines often drift or are erratic, especially for UV detection at lower wavelengths.

In this chapter, we will show that previous problems with gradient elution have been largely overcome by recent advances in equipment, materials, and a better understanding of the technique. In fact, many routine laboratories now make use of this important technique. Therefore, every laboratory involved in HPLC method development should have access to gradient elution, both for application to the routine separation of appropriate samples, and for developing isocratic methods. For a detailed discussion of the technique of gradient elution and the principles of gradient separations, see Refs. (1–4).

The following discussion is oriented mainly to reversed-phase HPLC. However, the application of gradient elution for other HPLC methods is quite similar (1–4). See Sects. 4.4 (ion-pair) and 4.5 (normal-phase) for the analogous procedures using isocratic separation.

6.1 APPLICATIONS OF GRADIENT ELUTION

Gradient Elution for Routine Analysis

Gradient elution is ideal for separating certain kinds of samples. These samples cannot be easily handled by isocratic methods, because of their wide k' range. A good example is shown in Fig. 6.1 for the anion-exchange separation of a mixture of aromatic carboxylic acids. In Fig. 6.1a, the isocratic separation of this mixture is attempted with a mobile phase containing 0.055 M sodium nitrate. Early eluting bands are poorly resolved (the first two peaks contain seven compounds), later peaks are too wide (and exhibit tailing), and

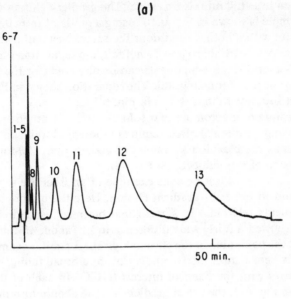

(a)

6-7

1-5

9

8

10

11

12

13

50 min.

(b)

12

11 13

9

4,5

3

2

10

6 7

25 min.

Figure 6.1 Separation of a Mixture of Aromatic Carboxylic Acids by Ion-Exchange Chromatography. (a) Isocratic separation with 0.055 M sodium nitrate in mobile phase. (b) gradient elution with sodium nitrate varying from 0.01 to 0.10 M. Reprinted with permission from Ref. (1).

the separation time (50 min) is excessive. The gradient-elution separation of this same sample is shown in Fig. 6.1b, using a gradient from 0.01 to 0.10 M sodium nitrate with all other conditions the same. Now, all peaks are better resolved, later peaks are sharp and do not tail, the separation time is reduced by half, and several trace components can be observed (for the first time) in the latter part of the chromatogram. The separation shown in Fig. 6.1b is in every respect superior to that shown in Fig. 6.1a.

The separation of macromolecular solutes such as synthetic polymers and biopolymers (e.g., proteins) often requires gradient elution. In many cases, samples such as these exhibit k' ranges of greater than 100-fold. For a detailed discussion of this subject, see Ref. (4).

Figure 6.2a shows a less obvious example of an isocratic separation that would be handled better by gradient elution. Here the quantitation of three active ingredients (steroids) in a medicinal ointment is desired, but these compounds showed a fairly wide difference in k' (about 15-fold), and poor resolution at the beginning of the chromatogram. A similar example is shown in Fig. 6.2b, where sample interferences elute as a broad tailing band near t_0, which interferes with the band of interest (HCT). In each of the two chromatograms in Fig. 6.2, the use of gradient elution should improve resolution and simplify the development of a final separation.

Other samples may separate well by isocratic elution, but the sample may also contain late-eluting compounds that can lengthen run time, result in gradual deactivation of the column, and/or interfere with the analysis of the next sample. Figure 6.3a shows an isocratic separation where the quantitation of a single compound (shaded band) is desired, but where later bands continue to elute indefinitely. Gradient elution would solve this problem. Figure 6.3b shows the gradient separation of a wood-pulp extract for the determination of anthraquinone. A broad, well-retained band (arrow) did not elute under isocratic conditions. The initial use of isocratic elution for these wood-pulp extracts resulted in a rapid loss of column activity and inadequate separation, due to the buildup of stongly retained compounds on the column. This problem was solved by changing to a gradient method.

Gradient elution is not applicable to every situation. The use of strongly retained additives in the mobile phase (e.g., triethylamine to prevent silanol interactions) can complicate the use of gradient elution, because column regeneration is slow and separations tend to be less reproducible. The same can be true of ion-pair chromatography, when the ion-pair agent is strongly retained by the stationary phase. Separations on bare silica are subject to similar problems, because polar solvents generally bind strongly to the sorbent. Gradient elution is best suited for reversed-phase systems, for normal-phase separations with polar bonded phases, and for ion-exchange HPLC.

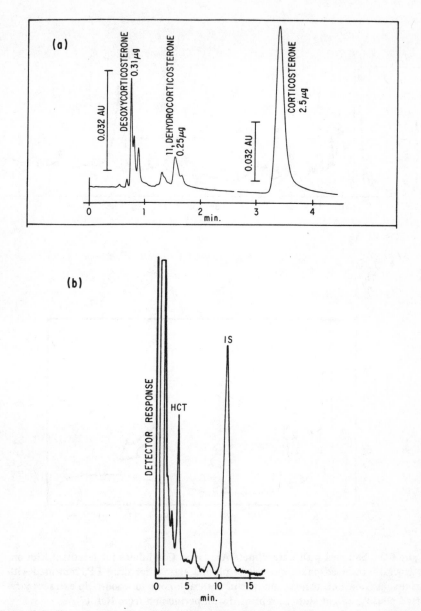

Figure 6.2 Samples that Benefit from Gradient Elution. (a) Dosage form containing various steroids as active ingredients; separation on silica column with methylene chloride as mobile phase, reprinted with permission from Ref. (5); (b) serum extract spiked with HCT (hydrochlorothiazide, a diuretic); reversed-phase separation with 15% methanol/water as mobile phase, reprinted with permission from Ref. (1).

(a)

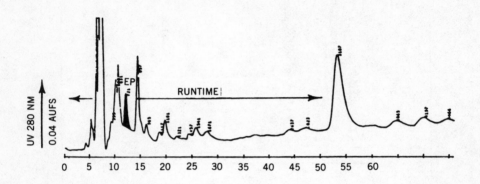

(b)

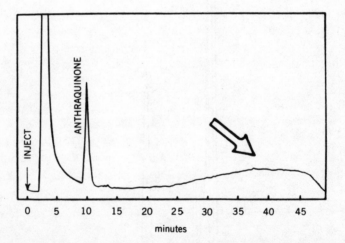

Figure 6.3 Samples with Late-Eluters Are Good Candidates for Gradient Elution. (a) Isocratic reversed-phase analysis of plasma extract for drug EP, reprinted with permission from Ref. (6); (b) analysis of anthraquinone in wood-pulp extract by reversed-phase gradient elution, reprinted with permission from Ref. (7).

158

Gradient Elution for Method Development

Section 2.2 (Eqn. 2.4) discusses the use of a single gradient run to infer which mobile-phase composition (e.g., what percent organic) would provide a suitable solvent strength for the isocratic separation of a given sample. An initial methanol/water gradient can be used to estimate the percent methanol required for suitable isocratic k'-values ($1 < k' < 20$). However, running two initial gradients allows the calculation of precise retention for each sample in isocratic elution, for any chosen percent organic value. Not only do these two gradient runs more precisely define the dependence of k' and run time on mobile-phase composition; they also create the possibility of identifying changes in band spacing as a result of small changes in percent organic (Sect. 4.2). To take full advantage of this procedure, a personal computer plus appropriate software is required, as discussed more fully in Chapter 8.

Using gradient elution to develop isocratic HPLC methods has a number of advantages, compared with using isocratic experiments. First, there are fewer trial-and-error adjustments in solvent strength required when changing from one solvent to another. Second, the ability to increase resolution during early exploratory runs is a distinct advantage when doing solvent mapping studies as in Sect. 4.3. Early bands often are severely overlapped in isocratic separation, so that it may not be clear how resolution is changing as separation conditions are varied. Gradient elution opens up the front of the chromatogram, allowing a better view of what is happening as conditions are varied.

Third, using gradient-elution runs during the initial stages of method development makes it easier to locate compounds that elute either very early or very late in the chromatogram. With isocratic separation, early-eluting compounds are often lost in the solvent front, while late-eluting compounds disappear into the baseline or overlap the next sample. Finally, gradient-elution method development works for either gradient-elution or isocratic methods. If the sample requires gradient elution in the final method, then no time will be wasted in carrying out initial isocratic separations that are obviously inadequate (and often time consuming). For these reasons, more workers are using gradient elution during HPLC method development, even when it is expected that the final method will be carried out by isocratic elution.

6.2   PRINCIPLES OF GRADIENT SEPARATION

Most chromatographers believe that gradient-elution separations are more complicated to perform and more difficult to understand than are isocratic separations. Perhaps as a result of this perception, few workers use a systematic approach for developing a final gradient-elution procedure. Actually, gradient elution can be easier to understand and use than isocratic elution. In any case, once a good understanding of isocratic separation is developed, a corresponding understanding of gradient elution is easily attained. The design of gradient procedures is similar to that for isocratic separations because these two elution techniques are based on the same fundamental processes of HPLC retention and separation. Once the dependence of k′ on mobile-phase composition has been defined, it is possible to predict retention in either isocratic or gradient elution (4,8).

Average k′ Value during Gradient Elution ($\bar{k}$)

Figure 6.4a portrays the migration of three solute bands (X, Y, Z) as they move through the column during gradient elution. The solid lines mark the fractional distance migrated (r) by each compound, between the column inlet and outlet at any time t during the separation (t = 0 marks the start of the gradient and the injection of the sample). The dashed lines indicate the instantaneous k′-value of each band at time t; k′ decreases with time, due to the increase in %-B and solvent strength during gradient elution. Resolution in gradient elution when using a linear gradient is determined by almost the same relationship (Eqn. 2.3) as for isocratic elution:

$$R_s = (1/4)(\alpha-1)N^{0.5}\,[\bar{k}/(\bar{k} + 1)]. \qquad (6.1)$$

The only difference in Eqn. 6.1 is that the average value of k′ during gradient elution ($\bar{k}$) replaces the isocratic quantity k′ of Eqn. 2.3. This average capacity factor $\bar{k}$ in gradient elution is defined in Fig. 6.4b; it is the instantaneous k′-value for a solute that has migrated halfway along the column. The following discussion applies quantitatively for the case of linear gradients and reversed-phase HPLC. The conclusions reached are approximately correct for other situations, involving nonlinear gradients and other HPLC methods. A further discussion of this is contained in Refs. (4,8).

The parameter $\bar{k}$ can be related to the conditions used in a gradient separation:

$$\bar{k} = (t_G F/1.15 V_m \Delta\Phi S). \qquad (6.2)$$

Here t_G is the gradient time (duration in min), F is flow rate (mL/min), V_m is the column dead volume (equal to [$t_0 F$]), $\Delta\Phi$ is the change in the volume fraction of the strong solvent during the gradient, and S is an isocratic parameter determined by the strong solvent and the sample compound. Plots of log k′ (isocratic) vs percent organic are usually linear, with a slope equal to S/100.

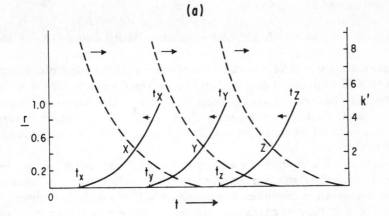

(a)

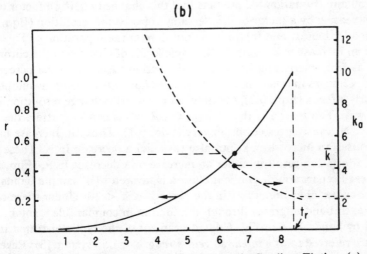

(b)

Figure 6.4 Migration of Bands through the Column in Gradient Elution. (a) migration of compounds X, Y, and Z; ——, fractional migration (r) between column inlet and outlet; ——, instantaneous k'-value during migration; (b) values of k' at column midpoint ($\bar{k}$). Reprinted with permission from Ref. (1).

The quantity $\Delta\Phi$ is given by

$$\Delta\Phi = [(\text{final } \%\text{-B}) - (\text{starting } \%\text{-B})]/100. \qquad (6.3)$$

For example, $\Delta\Phi = 0.5$ for a gradient that starts at 20% methanol/water and ends at 70% methanol/water. Values of S for small molecules (MW < 500) in reversed-phase HPLC usually range from 3 to 5. Differences in S for adjacent bands lead to band-spacing changes as percent organic is varied (see Sect. 4.3). Large molecules such as proteins can have S-values as large as 50–100.

Equation 6.2 is essential for understanding what happens during gradient separations. Since resolution depends strongly on $\bar{k}$ (Eqn. 6.1), it is important to select separation conditions to maintain $\bar{k}$ within a desirable range; typically, $2 < \bar{k} < 10$. Any change in gradient conditions (such as gradient time or flow rate) that affects $\bar{k}$ leads to predictable changes (Eqn. 6.2) in the chromatogram. This effect is illustrated in Fig. 6.5 for the same ion-exchange separation of Fig. 6.1. In Fig. 6.5a, the gradient time equals 20 min; in Fig. 6.5b, 50 min. Equation 6.2 predicts that this change in t_G (by a factor of 2.5) will increase $\bar{k}$ by a factor of 2.5, leading to (a) better resolution (Eqn. 6.1), (b) broader bands, and (c) a longer run time in the separation of Fig. 6.5b. Thus, an increase in $\bar{k}$ results in the same kinds of changes in the chromatogram as from increasing k' in isocratic elution (decreasing the percent organic). Changes in other conditions (such as flow rate or column volume) will similarly affect $\bar{k}$ (Eqn. 6.2), but may not necessarily change resolution in the same way. The reason is that changes in F or V_m can also affect the plate number N, causing resolution to vary (Eqn. 6.1). Thus, an increase in flow rate causes an increase in $\bar{k}$, but also (usually) a decrease in N. These combined effects can result in either an increase or a decrease in R_s (Eqn. 6.1).

Overall separation in gradient elution is favored if all sample bands have similar values of $\bar{k}$, preferably in the range $2 < \bar{k} < 10$. Under these conditions, each band migrates through the column in comparable fashion, illustrated by bands X–Z in Fig. 6.4a. A gradient of the type illustrated in this figure is referred to as a *linear-solvent-strength* or LSS system (3). Reversed-phase separations with linear gradients normally exhibit LSS behavior. For an (average) optimum value of $\bar{k} \cong 5$, and $S \cong 4$ (small molecule), Eqn. 6.2 becomes:

(optimum gradient in reversed-phase)
$$(t_G F/V_m \, \Delta\Phi) \cong 20. \qquad (6.4)$$

or,

$$t_G \cong 20 V_m \, \Delta\Phi/F. \qquad (6.4a)$$

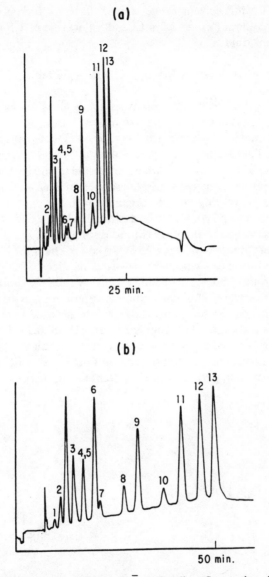

Figure 6.5   Effect of Gradient Time and $\bar{k}$ on Gradient Separation. Same separation as in Fig. 6.1. (a) Gradient time t_G equal to 20 min; (b) Gradient time t_G equal to 50 min. Reprinted with permission from Ref. (1).

For example, if a full solvent range is used (0–100% B, $\Delta\Phi = 1.0$) with a 25×0.46-cm column ($V_m \cong 2.5$ mL) and a flow rate of 1.5 mL/min, then the best gradient time is

$$t_G \cong (20 \times 2.5 \times 1.0)/1.5 = 33 \text{ min.}$$

Gradient times longer than this often yield only marginal increases in resolution, at the expense of wider (hard-to-detect) bands and long run times. However, large-molecule samples (mol. wt. $\geqslant 500$ Da) usually require longer gradient times, because of the larger S-values of the sample components.

It is often useful to vary $\bar{k}$ somewhat, to change band spacing and α, similar to the case for isocratic separation described in Sect. 4.2 (i.e., change in solvent strength by varying percent organic). In other cases (e.g., following the optimization of $\bar{k}$ and α), it may be desirable to hold $\bar{k}$ constant while other conditions are changed to increase N (and resolution). Figure 6.6 illustrates the possibility of varying band spacing as a result of changing $\bar{k}$, for the gradient elution separation of a complex mixture of peptides. In Fig. 6.6a the entire chromatogram is shown for two different flow rates, 0.5 and 1.5 mL/min. Figure 6.6b is an early part of each chromatogram, magnified to better view the individual bands. Because only the flow rate is varied between these two runs, values of $\bar{k}$ are larger at the higher flow rate (Eqn. 6.2). This change in $\bar{k}$ (gradient elution) caused by a change in flow rate, is equivalent to a change in k' caused by varying the percent organic solvent in isocratic separation. In many cases a change in percent organic solvent (isocratic runs) results in sub-

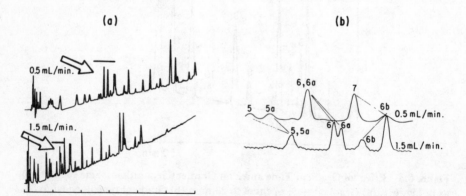

Figure 6.6 Effect of Flow Rate (and $\bar{k}$) on Band Spacing. Separation of peptides from enzymatic hydrolysis of myoglobin by reversed-phase gradient elution. (a) Whole chromatograms (flow rates of 0.5 and 1.5 mL/min); (b) partial chromatograms, showing change in α-values. Reprinted with permission from Ref. (10).

stantial band rearrangements (change in α-values); this is also the case in Fig. 6.6. When flow rate is increased to 1.5 mL/min, bands 5 and 5a are fused, bands 6 and 6a become partly resolved, and bands 6b and 7 change places. With a higher flow rate (<1.5 mL/min), all six bands can be resolved.

It is sometimes desirable to maintain the same (optimum) band spacing in a gradient separation, while changing column length, particle size, and/or flow rate to increase N or decrease run time. This requires that $\bar{k}$ be held constant by maintaining ($t_G F/V_m \Delta\Phi$) constant (Eqn. 6.2; S does not change for the same sample compound and same mobile-phase solvents).

Gradient Range

Apart from varying gradient conditions to achieve optimum values of $\bar{k}$ and α, it is important to adjust the gradient range (initial and final values of percent organic) so that (a) bands of interest do not elute too close to the beginning of the chromatogram (i.e., $k' \cong 0$), (b) all sample bands leave the column before the gradient is over, and (c) there is no wasted space at the beginning or end of the chromatogram. When adjusting the gradient range it is also important to keep in mind the dependence of $\bar{k}$ on $\Delta\Phi$ (Eqn. 6.2). These principles are illustrated below with some examples.

Figure 6.7a shows the initial separation of a mixture of substituted aromatic compounds, using a 5–95% acetonitrile/water gradient with a gradient time of $t_G = 20$ min. The beginning of the chromatogram is relatively empty, suggesting that the gradient can be started with a higher percent organic; 46% acetonitrile/water was chosen in Fig. 6.7b. The $\bar{k}$ is maintained constant for this second run, and $\Delta\Phi$ is reduced from 0.90 (95%–5%) to 0.49 (95–46%), so the gradient time must be reduced proportionally (Eqn. 6.2). Therefore, $t_G = 12$ min for the 46–95% acetonitrile/water gradient of Fig. 6.7b, vs. 20 min in Fig. 6.7a. The resulting separation in Fig. 6.7b differs from that in Fig. 6.7a only in a compression of the chromatogram front. Since this region was initially uncrowded, the separation in (b) is not inferior to that in (a); however, 40% of the original run time has been eliminated. Note the virtually identical form of each chromatogram (Figs. 6.7a and b) after band 4; this pattern is typical of separations where the starting percent organic is changed, but $\bar{k}$ is kept constant by varying t_G in proportion to $\Delta\Phi$. However, one objection to the run of Fig. 6.7b is that the first band elutes near t_0; this can be undesirable in gradient elution, just as in isocratic separation.

In Fig. 6.7c the gradient range was reduced again (79–95% acetonitrile), with a further reduction in run time ($t_G = 4$ min). Now, the resolution of the first eight bands is inadequate. From this example, it is apparent that there is an optimum gradient range for every sample (in this case, the one in Fig. 6.7b). An optimum gradient range will usually position the first band at

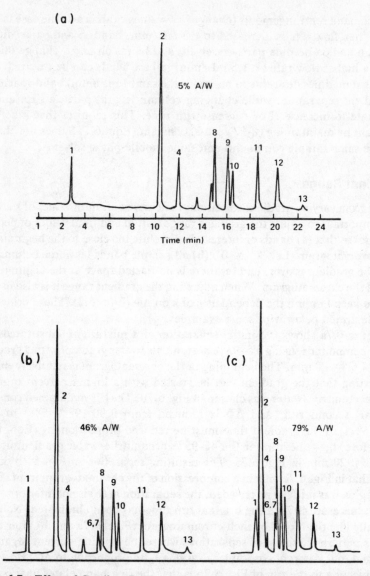

Figure 6.7 Effect of Gradient Range on Separation. Mixture of substituted aromatics; reversed-phase conditions and acetonitrile/water gradients. (a) 5–95%B, $t_G = 20$ min; (b) 46–95%B, $t_G = 12$ min; (c) 79–95%B, $t_G = 4$ min. Reprinted with permission from Ref. (3).

about $2t_0$, and the last band at the end of the gradient. An additional example is given in Figs. 6.8a,b, described in the following section.

6.3   OPTIMIZING GRADIENT METHODS

Since gradient-elution separations are similar to isocratic runs, it follows that a similar approach to method development is possible. The same conditions recommended for initial isocratic separations can be used (see Table 1.2). The next step (see Sect. 4.2) is to select an appropriate percent organic so as to provide k'-values in the range of $1 < k' < 20$. For an isocratic separation this is done empirically by trying different mobile-phase compositions (e.g., Fig. 4.4). In the case of gradient elution, conditions can be predicted in advance of the first separation, so that values of $\bar{k} \cong 5$ for all bands (Eqn. 6.4). Thus, the first gradient separation should provide $\bar{k}$-values that are close to optimum ($\bar{k} \cong 5$) for all sample bands. Once this is accomplished, resolution and run time can be systematically improved by varying values of α and N (Eqn. 6.1). This method-development approach is now illustrated with a typical sample: a mixture of peptides from the enzymatic hydrolysis of lysozyme (11). Table 6.1 summarizes the steps in developing a final gradient separation.

Optimizing the Gradient Range

The initial gradient separation of the peptide digest from the protein lysozyme is shown in Fig. 6.8a. On the basis of experience with similar samples, the initial gradient range was selected as 5–60% acetonitrile/water. The starting experimental conditions involved a 8-cm-long column, a flow rate of 1.9 mL/min, and a gradient time of 30 min (this is greater than predicted by Eqn. 6.4, because $S \cong 10$ for this higher-molecular-weight sample). Note that the first major bands do not appear until about 5 min after t_0, while the last band leaves the column 16 min after t_0. To save run time, the gradient can be started with a higher percent organic, and completed with a lower percent organic, without affecting resolution (cf. Fig. 6.7).

Estimates of the preferred gradient range were obtained from Fig. 6.8a as follows. It was decided to eliminate that part of the gradient corresponding to 0–4 min and 16–30 min. The original gradient (Fig. 6.8a) was 5–60% in 30 min, or 1.8%/min. The second gradient was therefore started at $(5 + 1.8 \times 4) = 12\%$ and terminated at $[(16 - 4) \times 1.8] + 12 = 34\%$, in a time t_G equal to $(16 - 4) = 12$ min. The separation shown in Fig. 6.8b is for a 12–34 %-B gradient in a time of 12 min. A comparison of these two runs (Figs.

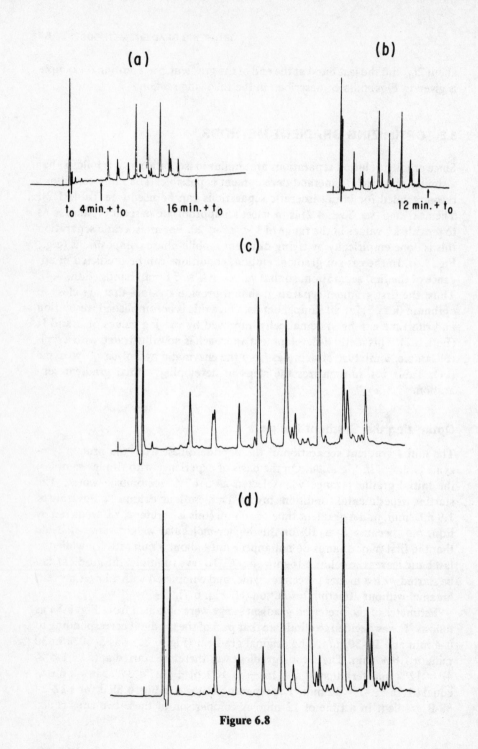

Figure 6.8

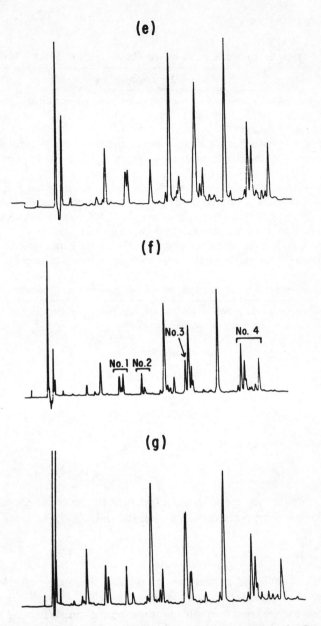

Figure 6.8 Optimizing the Reversed-Phase Gradient Separation of a Peptide Mixture (Lysozyme Digest). (a) initial separation, 5–60% acetonitrile/water; 8.0 × 0.62-cm, 5-μm C-8 column; 1.9 mL/min; $t_G = 30$ min; 35° C; (b) same as (a) except 12–34%B in 12 min; (c) same as (b), except $t_G = 15$ min, 2 mL/min; (d) same as (b), except $t_G = 30$ min, 1 mL/min; (e) same as (b), except $t_G = 60$ min, 0.5 mL/min; (f) same as (b), except two columns in series (16 cm), $t_G = 60$ min, F = 1 mL/min; (g) same as (e), except 2.5 mL/min. Reprinted with permission from Ref. (11).

TABLE 6.1   Steps in Optimizing a Reversed-Phase Gradient-Elution Method

Step	Description
1	Select standard gradient conditions for first run; use conditions given in Table 1.2: 5–100 % acetonitrile/water; gradient time of 25 min (2 mL/min) to 50 min (1 mL/min).[a]
2	Adjust the gradient range to minimize wasted time (empty space) at beginning or end of the chromatogram (see Fig. 6.7); change t_G in proportion to $\Delta\Phi$.
3	If the chromatogram is quite complex at this point (many bands, poor resolution), increase N by increasing t_G or column length, and/or by decreasing particle size and flow rate; keep $\bar{k}$ (or $t_G F/V_m$) constant while adjusting t_G, flow rate, and column dimensions (Eqns. 6.2 and 6.4).
4	Once there is room in the chromatogram for all the bands (and not too many bands are overlapped), adjust band spacing as necessary by varying flow rate.
5	For further changes in band spacing, change the organic solvent.

[a]If the sample has a mol. wt. > 500, revise the initial gradient conditions (Eqn. 6.4); increase t_G in proportion to S, where S can be estimated as follows:

Sample Mol. Wt.	S
100	4
1,000	10
10,000	30
100,000	100

For THF as solvent, increase t_G values by a factor of 1.5.

6.8a,b) shows that they are indeed equivalent with respect to resolution, but the second separation is completed in half the time.

Optimizing N

At this point, a choice of approaches is available. For many samples, where the number of components is not too large (less than 15 peaks), changes in band spacing can be tried to improve sample resolution. When the sample is more complex (e.g., the peptide sample of Fig. 6.8b), it often is preferable to first increase N for the improved resolution of all bands. This approach permits a better view of what is happening as changes in band spacing are attempted. It also provides additional space in the chromatogram for bands that may separate when band spacing is altered. For less complex samples, the column conditions (and the value of N) can be optimized *following* changes in band spacing.

Figure 6.8c is essentially the same as Fig. 6.8b (t_G = 15 min and flow rate = 2 mL/min, vs. 12 min and 1.9 mL/min in Fig. 6.8b). The column plate count can be increased by reducing flow rate and/or increasing column length, just as in isocratic separation (see Sect. 2.5). Since $\bar{k}$ was initially optimized by the choice of gradient conditions, it should be kept constant at this stage in method development. This is analogous to keeping mobile-phase composition constant in isocratic separation, while increasing N and resolution via changes in flow rate or column length.

If the flow rate is decreased twofold, t_G must be increased by the same factor to hold $\bar{k}$ constant (see Eqn. 6.4). Figures 6.8d,e show the results of successively decreasing flow rate and increasing t_G: (d) 1 mL/min and t_G = 30 min and (e) 0.5 mL/min and t_G = 60 min. Some improvement in resolution is seen for the run of Figs. 6.8e vs. 6.8c, but a significant increase in resolution cannot be expected as a result of further decreases in flow rate. The same situation arises in isocratic separations, where optimum flow rates (see Sect. 2.5) for maximum N occur in the range of 0.3–1 mL/min for small-particle column packings (0.46 cm i.d. columns).

N can also be increased by increasing column length. Again, other gradient conditions must also be adjusted accordingly (Eqn. 6.4); there must be proportional increases in either t_G or flow rate. Figure 6.8f shows the result of doubling column length (16 cm) while increasing flow rate from 0.5 to 1 mL/min, and holding t_G constant (this keeps $\bar{k}$ constant); a significant further increase in resolution is obtained. This approach could be continued, if necessary, by further increasing column length (and increasing t_G and/or flow rate proportionally). However, Fig. 6.8f now shows enough empty space in the chromatogram to attempt further increases in resolution by changing band spacing.

Optimizing Band Spacing: Change in $\bar{k}$

Values of α can be adjusted in gradient elution in the same manner as for isocratic separations (Sect. 4.3). This approach is either to change $\bar{k}$ (by varying t_G or F in gradient elution), which is equivalent to a change in k' (by varying percent organic in isocratic separation), or to try a different organic solvent. A change in $\bar{k}$ is simpler experimentally, and should be tried first. The potential power of a change in $\bar{k}$ (by changing flow rate) for a different peptide sample was illustrated in Fig. 6.6. Similar band-spacing changes for the lysozyme digest are shown in Fig. 6.8, where the flow rate was changed from 0.5 mL/min in Fig. 6.8e to 2.5 mL/min in Figure 6.8g (other conditions held constant). Close examination of these two runs shows numerous changes in band spacing that could be the basis of an improved separation. For samples as complex as this, it will usually be necessary to try several flow-rate changes to obtain the best overall separation (see discussion of Ref. 10). See also the further discussion of Sect. 8.4.

Optimizing Band Spacing: Change in Organic Solvent

To illustrate this approach for optimizing band spacing in gradient elution, a 14-component mixture of synthetic organic compounds will be used (12). The initial gradient run in this case is shown in Fig. 6.9a, a 20–80% methanol/water gradient; the gradient range has already been optimized to eliminate wasted time at the beginning and end of the chromatogram. Twelve of the 14 compounds are separated, and there is enough space in the chromatogram to take advantage of any band-spacing changes that can be achieved. Methanol in the mobile phase was first replaced by acetonitrile (Fig. 6.9b) and then tetrahydrofuran (Fig. 6.9c). Neither of these latter gradient chromatograms is superior to that of Fig. 6.9a, but a solvent-optimization strategy was continued according to the standard (isocratic) scheme of Figs. 4.5 and 4.6. Comparison of the resulting seven gradient runs suggested the optimum final gradient shown in Fig. 6.9d. The separation of this sample by the latter gradient is shown in Fig. 6.9e, where all 14 compounds are resolved with $R_s > 1.3$.

The preceding examples demonstrate that method development in gradient elution can proceed in much the same way as in isocratic elution. Nevertheless, additional separation variables (gradient time and gradient range) are involved, so that the overall process of method development is necessarily more complicated. Furthermore, the dependence of $\bar{k}$ on such variables as column length, flow rate, and gradient time must be kept in mind as experimental conditions are varied for improved separation. For these and other reasons, the use of a computer-assisted method-development approach is more useful for gradient methods than for isocratic separation (see Chapt. 8).

6.4 EXPERIMENTAL CONDITIONS

At the beginning of this chapter, several experimental problems encountered in gradient elution were noted. However, most of these problems can be resolved by taking appropriate action.

Reproducible Separations: Column Equilibration

Many workers believe that gradient elution is less reproducible than isocratic separation. When gradient runs cannot be duplicated from run to run or be-

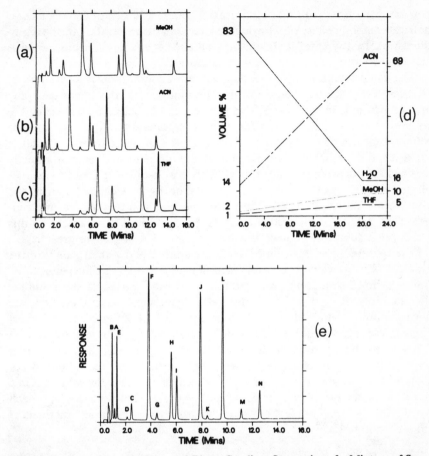

Figure 6.9   Optimizing the Reversed-Phase Gradient Separation of a Mixture of Synthetic Organic Compounds; Band-Spacing Changes via a Change in Organic Solvent. (a) Methanol/water gradient; (b) acetonitrile/water gradient; (c) THF/water gradient; (d) optimum final gradient (four solvents); (e) separation with gradient of (d). Reprinted with permission from Ref. (12).

tween different laboratories, one of two possibilities should be considered. The most likely problem is that the column is not properly regenerated between runs. In gradient elution, it is important that the column be completely equilibrated with the starting mobile phase before injecting the next sample and beginning the gradient. Column equilibration usually can be achieved by flushing the column with 15–20 column volumes of the starting mobile phase. For example, a 25 × 0.46-cm column (V_m = 2.5 mL) and a 5–100% methanol/water gradient requires that at least 37 mL (15 $\times$ 2.5) of the 5% metha-nol/water mobile phase flow through the column before starting the next gradient run. In any case, it should be verified that the equilibrium period is sufficient to give reproducible chromatograms.

The usual symptom of inadequate column equilibration is a variation in retention times for early bands in the chromatogram; later bands generally are not affected. In most cases, this problem can be solved by increasing the volume of mobile phase used for column regeneration. If the volume of column-regeneration mobile phase required for constant retention times is excessive, it is probable that some component of the mobile phase is strongly retained (see discussion of Sect. 6.1). This may require adding the mobile-phase additive to both the A and B solvents, or (in some cases) removing the additive altogether. Long equilibration times are also required in reversed-phase separations, if the gradient is started with water (0% organic). For this reason, it is advisable to begin the gradient with 5% organic or greater.

The time required for column regeneration is usually about equal to the gradient time t_G, which can mean a doubling of the analysis time per sample. This additional time can be reduced in various ways. Since column equilibration is mainly affected by the volume of the regeneration solvent, the flow rate during column equilibration can be increased. It is also possible to stop short of complete column equilibration (with a reduction in both time and solvent volume), providing that samples are injected using an autosampler, so that the time between injections is kept constant. In this case, the column is not completely equilibrated at the time of sample injection, but the column is in a constant, semiequilibrated state for each sample. However, this approach should only be used as a last resort, when other more traditional methods of equilibration fail.

Reproducible Separations: Equipment Differences

All gradient systems are not equivalent. Well-designed equipment will provide gradients that conform closely to the expected shape; e.g., linear as in Fig. 6.10a. However, significant differences in the gradients provided by different instruments are not uncommon. For this reason, it is important to measure the actual gradient provided by each HPLC system as follows:

1. Remove the column from the system, and connect the sample injector directly to the detector with a minimum length of 0.25-mm i.d. capillary tubing. (Note: it may be necessary to add a restriction after the detector for the pump(s) to work properly.)

2. Use methanol as the solvent for both gradient pumps (A and B solvents). Add a UV-absorbing compound (e.g., 0.2%v acetone) to both

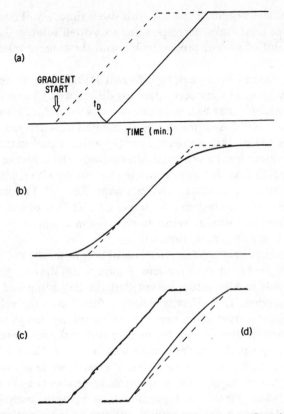

Figure 6.10   Gradient Profiles for Different HPLC Equipment. Standard conditions; $t_G = 20$ min, $F = 2$ mL/min, linear gradient from 0–100% B. Reprinted with permission from Ref. (13).

solvent bottles, in sufficient quantity to produce a 10% deflection from baseline for the A solvent and a 90% deflection for the B solvent. If the detector wavelength can be varied, this can be used to fine-tune the absorbance of the B solvent. For these experiments, the detector range must be below the maximum limit of linear response for the detector.

3. Run a linear gradient from 0 to 100% B with $t_G = 20$ min and a flow rate of 2 mL/min.

The resulting display on a recorder stripchart shows the gradient as a function of time. A typical result is illustrated in Fig. 6.10a. The arrow shows the time at which the gradient was started, and the solid curve is the observed gradient profile. The actual gradient (solid line) is displaced by a time t_D, the dwell time of the gradient system. Typically, this dwell time t_D will equal 1–3 min for a flow rate of 2 mL/min, corresponding to a dwell volume V_D of 2–6 mL between the point of solvent proportioning and the column inlet (note that $t_D = V_D/F$).

If the dwell time t_D varies among different HPLC systems, the resulting gradient separations can show corresponding differences. Changed resolution for early eluting bands may be a symptom of a change in t_D. Differences in t_D between different HPLC systems can be eliminated by a delayed injection of sample, relative to the start of the gradient (by a time equal to the difference between the t_D values for the systems). Alternatively, the choice of equipment for a particular HPLC assay can be restricted to systems with similar values of t_D. In high-pressure gradient systems (see Chapt. 3 of Ref. 1) usually $V_D \cong 2$ mL, while in low-pressure systems $V_D \cong 5$–7 mL. The use of equipment with different t_D-values will usually result in changes in retention time, but few other changes in the chromatogram.

Gradient equipment also differs in terms of how faithfully the desired gradient is actually produced by the system. Figure 6.10b shows a gradient profile similar to that in Fig. 6.10a, except that the beginning and end of the gradient are rounded. This effect is usually a function of the volume of the gradient mixer, pulse dampener, and connecting tubing. Large-volume mixers (usually found with low-pressure mixing systems) give more rounding, other factors being equal. Gradient profiles as in Figs. 6.10c,d reflect a poor gradient system. In Fig. 6.10c the gradient is erratic, while in Fig. 6.10d the gradient is not linear. Equipment that produces gradients as in these latter two examples, is less suitable for reproducible and precise analysis by gradient elution; system-to-system variability will also be more noticeable. However, most equipment sold today is not subject to these problems.

Baseline Problems

In isocratic separations, it is usually possible to obtain a baseline (no sample injection) that is flat (no drift), without false peaks or other disturbances. Such a baseline can be more difficult to achieve in gradient elution for various reasons. Therefore, it is imperative to run a "blank gradient" before injecting samples for gradient separation; that is, run a gradient without injecting a sample. An example is shown in Fig. 6.11, where distilled water was used in the mobile phase in the top chromatogram (0–100% acetonitrile/water, blank gradient) vs. the use of HPLC-grade water in the bottom chromatogram. Numerous artifactual bands are apparent in the top gradient, due to impurities in the distilled water used for this run. Obviously, a situation like this will preclude (or greatly complicate) the successful development of a gradient-elution procedure. However, gradient artifacts as observed in Fig. 6.11

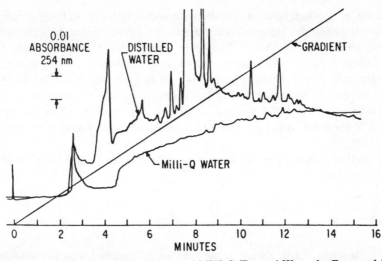

Figure 6.11 Comparison of Distilled and Milli-Q-Treated Water by Reversed-Phase Gradient Elution. Column, 30 × 0.4-cm μBondapak C-18; mobile phase, linear acetonitrile/water gradient; flow rate, 4 mL/min; detector, UV, 254 nm; 0.1 AUFS. Reprinted with permission from Ref. (14).

can always be seen by setting the detector for high sensitivity and a sufficiently low wavelength. Therefore, the blank gradient should be run under conditions (detector settings and gradient range) that are the same as will be required for the actual gradient run with injection of sample.

Artifacts observed in a blank gradient can arise from impurities in either solvent (acetonitrile or water in the example of Fig. 6.11). The solvent used to dissolve the sample (if different from the A solvent) is also a possible source of impurities that end up as spurious peaks. The use of HPLC-grade solvents minimizes baseline problems, but this approach may not be enough when using a UV detector at wavelengths below 220 nm and high detector sensitivity. When problems of this kind are discovered, the best approach is usually to try other sources for the solvents. HPLC-grade solvents can vary significantly in impurities from one source to another, and even among different lots from the same supplier. Incomplete degassing of the mobile phase can also result in a broad band (small air bubbles) that elutes at about 50% organic.

In Fig. 6.11 the bottom (good) chromatogram does not provide a flat baseline. Some distortion of the baseline is common in all gradient-elution procedures, but this need not compromise the final method. Sometimes such distortions are due to refractive-index differences between the A and B solvents. UV detectors vary in their ability to ignore these effects, so different modules should be tried when baseline problems are encountered with a particular detector. Baseline drift in gradient elution can also be caused by differences in the absorbance of the A and B solvents. Water does not absorb above 180 nm, whereas most organic solvents do. Therefore, upward-drifting baselines are commonly seen for detection at low wavelengths, because the mobile-phase absorbance increases during the gradient (as %-B increases). If this problem interferes with the method, it can be eliminated by adding a nonretained UV absorber to the A solvent (e.g., salts or ionized buffers of various kinds) (15).

For further information on experimental problems in gradient elution, consult Ref. (16).

REFERENCES

1. L. R. Snyder and J. J. Kirkland, *Introduction to Modern Liquid Chromatography,* 2nd ed., Wiley-Interscience, New York, 1979, Chapt. 16.
2. P. Jandera and J. Churacek, *Gradient Elution in Column Liquid Chromatography,* Elsevier, Amsterdam, 1985.
3. L. R. Snyder, in *High-Performance Liquid Chromatography. Advances and Perspectives, Vol 1,* Cs. Horvath, ed., Academic Press, New York, p. 207.
4. L. R. Snyder and M. A. Stadalius, in *High-Performance Liquid Chromatography. Advances and Perspectives, Vol. 4,* Cs. Horvath, ed., Academic Press, New York, 1986, p. 195.
5. *The HPLC Applications Book, Vol. 1.* Hewlett Packard, 1974.
6. F. Erni, H. P. Keller, C. Morin, and M. Schmitt, *J. Chromatogr., 204* (1981) 65.
7. K. H. Nelson and D. Schram, *J. Chromatogr. Sci., 21* (1983) 218.
8. J. W. Dolan, J. R. Gant, and L. R. Snyder, *J. Chromatogr., 165* (1979) 31.
9. M. A. Quarry, R. L. Grob, and L. R. Snyder, *Anal. Chem., 58* (1986) 907.
10. J. L. Glajch, M. A. Quarry, J. F. Vasta, and L. R. Snyder, *Anal. Chem., 58* (1986) 280.
11. M. A. Stadalius, M. A. Quarry, and L. R. Snyder, *J. Chromatogr., 327* (1985) 93.
12. J. J. Kirkland and J. L. Glajch, *J. Chromatogr., 255* (1983) 27.
13. L. R. Snyder and J. W. Dolan, *DryLab G User's Manual,* LC Resources Inc., San Jose, Calif., 1987.
14. R. L. Sampson, in *Liquid Chromatography of Polymers and Related Materials,* J. Cazes, ed., Marcel Dekker, New York, 1976, p. 149.
15. Sj. van der Wal and L. R. Snyder, *J. Chromatogr., 255* (1983) 463.
16. J. W. Dolan and L. R. Snyder, *Troubleshooting HPLC Systems,* Humana Press, Clifton, N.J., 1989.

7

SPECIAL
SAMPLES AND
TECHNIQUES

Previous chapters have presented general procedures for HPLC method development. The special needs of some samples have been intentionally minimized in this approach. While it is our belief that most samples can be processed by the same approach, this is not true for some compounds. For example, if the objective is to separate chiral isomers, none of the techniques presented so far is capable of accomplishing this goal. Likewise, the separation of inorganic ions is favored by columns and operating conditions that are largely different from those used for other samples. In addition, if the separation of interest is a trace analysis, some modification in separation goals will be required. Finally, some of the less common HPLC methods have not yet been discussed. In this chapter we will deal briefly with various special methods.

7.1 INORGANIC IONS AND ION-EXCHANGE CHROMATOGRAPHY

Mixtures of inorganic ions are not the typical sample for most HPLC labs. However, such samples are often analyzed by the technique of ion chromatography, which is a special variation of ion-exchange chromatography (IEC). We will refer to ion chromatography only briefly, since detailed discussions are found in Refs. 1 and 2. Ion-exchange chromatography (IEC), on the other hand, is an important technique for separating both inorganic ions and ionizable organic compounds (e.g., separation of large biomolecules). Therefore, method development in IEC will be discussed in more detail.

Ion-Exchange Chromatography

IEC separations are similar to those carried out by ion-pair chromatography (IPC, Sect. 4.4). Compounds that are ionic or ionizable (acids and bases) often can be separated by either HPLC method. In IPC, the surface of the column packing is coated by the sorbed ion-pair agent, resulting in a net charge on the packing surface. The resulting surface acts very much like that in an IEC column, except that, in the case of IEC, the charged groups are *covalently* bound to the surface of the packing. Ionized species (acids or bases) are retained by displacing a counter-ion that is initially associated with the ionic group bound to the particle surface. The retention of a sample ion X^+ onto a cation-exchange column can be represented as:

$$X^+ + (S^- Na^+) \rightleftharpoons Na^+ + (S^- X^+). \tag{7.1}$$

Here, X^+ and Na^+ refer to ions in the mobile phase, with Na^+ being the counter-ion in this example; an anionic group on the particle surface is represented by S^-. Similarly, the retention of a sample ion X^- in an anion-exchange column (e.g., with NO_3^- as the counter-ion) is given as

$$X^- + (S^+ NO_3^-) \rightleftharpoons NO_3^- + (S^+ X^-) \tag{7.2}$$

From Eqns. 7.1 and 7.2, it is apparent that an increase in the mobile-phase counter-ion concentration (e.g., Na^+ or NO_3^-) will proportionately decrease the retention of the sample ion. In IEC, compounds having different charges (e.g., $+1$ or $+2$) will also show changes in α as the salt concentration is varied. Since a change in pH can affect the relative ionization of acids or bases, higher pH values lead to increased ionization and greater retention of acids on anion exchangers; lower pH values favor increased ionization and retention of bases on cation exchangers.

Table 7.1 summarizes the similarities and differences between ion-exchange and ion-pair HPLC. It has already been noted that, compared with IPC, IEC usually is favored for separating inorganic ions and large biomolecules. In some cases, IEC is also a good alternative for separating small organic molecules. While specific ion-exchange columns are required for IEC, any reversed-phase column can be used for ion-pair chromatography.

Consider next the selection of the mobile phase for each method. IEC is generally carried out with aqueous mobile phases, in contrast to the use of water/organic mixtures in IPC. Retention in IEC is usually controlled by varying the concentration of salt or buffer in the mobile phase; concentrated salt solutions are the equivalent of strong solvents, and dilute salt solutions are weak solvents. In IPC, the retention of ionizable solutes can be altered by changing pH and/or the type and concentration of ion-pair agent. However, solvent strength is usually varied in the same manner as for reversed-phase HPLC: by changing the percent organic. IEC is often used in a gradient elution mode, with salt gradients of 0–0.5 M salt; whereas, gradient elution can

TABLE 7.1 Similarities and Differences between Ion-Exchange and Ion-Pair Chromatography

Feature	Ion-Exchange	Ion-Pair
Samples	Any ionic sample, especially inorganic ions and large biomolecules	Any ionic sample
Column	Anion or cation exchange	Reversed-phase (C-18, C-8, etc.)
Mobile phase	Aqueous solution of buffer and/or salt	Water/organic plus buffer and ion-pair agent
Increase k'	Decrease salt	Decrease organic
Increase α	Vary pH, change salt type; e.g., use K^+ vs. Na^+	Vary pH or concentration of ion-pair agent
Possible problems	Greater column variability; less-stable columns; less control over band spacing	Slow equilibration of column when mobile phase is changed; problems with gradient elution

be difficult in IPC, due to potentially long equilibration times of the noncovalently bound ion pairs.

For both IEC and IPC, compounds with different pK_a values usually can be separated by varying the mobile-phase pH to change band spacing. In IEC, band spacing can be further altered by changing the salt concentration, varying the temperature, or adding small amounts (5–20%) of different organic solvents. In IPC, percent organic, pH, and the concentration of the ion-pair agent are the primary variables used for changing band spacing (see Sect. 4.4).

Problems can be encountered when using either IEC or IPC. Column-to-column variability, as discussed in Sect. 3.1, tends to be more of a problem when using bonded-phase ion-exchange packings. Similarly, these IEC columns can be less stable than corresponding reversed-phase columns. The ability to vary the concentration of charged groups on the surface of the column packing in ion-pair chromatography adds an important variable for controlling band spacing. This option is seldom available in IEC separations.

Ion-pair chromatography, on the other hand, is subject to different problems. Long equilibration times are sometimes required when changing the mobile phase in IPC, due to the slow removal of the ion-pair agent from the stationary phase. Problems with baseline upsets and artifacts in ion-pair HPLC are common, due to impure ion-pair agents. Finally, ion-pair chromatography is less well suited to gradient elution because of difficulties in maintaining equilibration. Therefore, samples with a wide k' range may be better separated using ion-exchange chromatography.

Inorganic Analysis

Ion chromatography generally is the preferred technique for assaying inorganic ions at low concentrations in aqueous solutions. This method is especially useful for anions, and for elements occurring in different oxidation states. In the usual form of ion chromatography, the separation is performed on an initial "separator" column using a dilute salt solution as mobile phase. This salt is subsequently removed by a second, high-capacity IEC column ("stripper" column), usually as a result of neutralization of the counter-ions. For example, in a cation assay, the mobile phase is usually a dilute acid solution and the stripper column is an anion-exchange column, often with a hydroxide counter-ion. Once the mobile phase salt is removed from the mobile phase by the stripper column, sample ions can be detected with high sensitivity (ppb range) using conductivity detection, a universal sensor for aqueous solutions of ionic species. References 1 and 2 provide further details for these separations.

7.2   SIZE-EXCLUSION CHROMATOGRAPHY

Characterization of Macromolecules

Size-exclusion or gel-permeation chromatography (SEC or GPC) is a standard method to obtain information about the molecular weights and molecular-weight distributions of synthetic polymers. The method is also widely used to perform separations of naturally occurring macromolecules. The sample solution is introduced into a column that is filled with a porous column packing and is carried by a mobile phase through the column. Separation by molecular size occurs by repeated exchange of the solute molecules between the bulk solvent of the mobile phase and the stagnant liquid within the pores of the packing. The pore size of the packing particles determines the molecular size range wherein separation can take place. Columns of large-pore packings separate higher-molecular-weight components; lower-molecular-weight materials are separated by columns of smaller-pore packings. With columns of the proper pore size, molecules in the $50-10^7$ molecular-weight range can be separated. In general, resolution is significantly poorer than that afforded by the other HPLC methods. Column packings are available for both organic and aqueous mobile phases.

Standards of the same molecular type are required for calibration to convert SEC chromatograms into molecular-weight distributions and to calculate molecular-weight averages. The method development and data-handling techniques of SEC are much too involved for discussion in this book. Therefore, we suggest that Refs. 3–5 be consulted for details of this important characterization method for macromolecules.

SEC is an effective first-analysis method for unknown mixtures. Using a column set that fractionates over a wide molecular weight range, SEC can quickly define the sample in regard to complexity and the presence and range of high and low molecular weight components. This important information usually can be gained in a single run with no method development (3).

SEC often can be very useful as a preliminary method for simplifying the analysis of very complex mixtures. This method provides a means of easily classifying the sample into fractions containing components with different mass (molecular weight). After prefractionation by SEC, the resulting fractions can then be more readily separated into individual components by another method, such as reversed-phase HPLC.

The SEC method also can be valuable for eliminating unwanted components from a sample, to simplify the measurement of one or a few compounds in a complex mixture. For example, by selecting SEC columns with the proper pore size, potentially interfering higher-molecular-weight components can be readily removed from a plant or animal extract, so that a small molecule of interest (e.g., a drug or pesticide) can be separated and measured by another HPLC method. For further discussion of the use of SEC as an effective prefractionation method, both in manual or on-line column-switching modes, see Refs. (6,7).

7.3 ENANTIOMERS

Within the past decade there has been a rapid increase in separations of enantiomeric compounds (optical or chiral isomers). Enantiomers are isomeric compounds that have one or more chiral (asymmetric) centers in the molecule—usually a carbon atom substituted with four different groups. Interest in these compounds stems from their importance to pharmaceutical, agricultural, and other related fields. In general, the biological activity of two chiral isomers is different. For example, in the case of the drug penicillamine, the D-isomer possesses useful therapeutic properties, while the L-isomer is highly toxic. Therefore, assays to determine the enantiomeric purity of penicillamine are required in the course of its production for use as a drug. Today, such assays for chiral purity are usually carried out by HPLC.

The HPLC separation of underivatized enantiomers cannot be achieved by the approaches described in preceding chapters. The reason is that chiral isomers possess the same physical and chemical properties and are, therefore, unresolved in most HPLC systems ($\alpha = 1.00$). Chiral separations require the formation of diastereomers or diastereomeric complexes from the original enantiomers, using an optically pure chiral reagent. This effect is illustrated in Fig. 7.1a for the reaction of reagent ($+$) (QRS)CJ with the enantiomers ($+, -$)(XYZ)CH. HJ is eliminated, resulting in a covalent bond between the two chiral carbon centers. The resulting diastereomers ($+,+$ and $+,-$) differ in their physicochemical properties and can be separated by conventional HPLC techniques, for example reversed-phase.

The covalent carbon–carbon bond in the diastereomers of Fig. 7.1a can be replaced by a physical attraction between the enantiomer and the chiral reagent. This is illustrated in the example of Figure 7.1b, for enantiomers ($+, -$)(S_1, S_2, S_3, S_4)C and reagent ($+$)(R_1, R_2, R_3, R_4)C. Groups S_1 and R_4 attract each other, as do groups S_2 and R_3. The resulting diastereomeric complex will exhibit different stabilities (and physicochemical properties) for the two sample enantiomers, leading to the possibility of their separation by HPLC.

The separation of diastereomers or diastereomeric complexes (Fig. 7.1) requires at least one further interaction (attractive or repulsive) of the various substituent groups around the two chiral centers, for example, the interaction of groups S_3 and R_2 in Fig. 7.1b. This means that the two chiral centers must

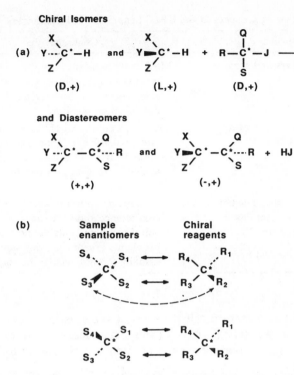

Figure 7.1 Illustration of (a) Enantiomers and Related Diastereomers Resulting from Reaction of Enantiomers with Optically Pure Chiral Reagent; (b) Physical Attraction between Enantiomers and Chiral Reagents. ◄──►, attractive (physical) interaction; ─ ─ ─, attractive or repulsive (physical) interaction.

be in reasonable proximity to each other. This third interaction (in the case of Fig. 7.1b) can be either attractive (e.g., hydrogen bonding) or repulsive (steric hindrance), but the strength of this interaction must be different for the two enantiomers. This is the case in the example of Fig. 7.1b, because of differences in the distance between groups S_2 and R_3 in each diastereomeric complex. Many chiral separations are based on diastereomeric complexes, rather than diastereomers. The need for three interactions in the complex (as shown in Fig. 7.1b) is referred to as the *Dalgleisch rule*.

The HPLC separation of chiral isomers can be achieved in three general ways, as summarized in Table 7.2. Most workers begin with chiral columns, which have the chiral reagent as part of the stationary phase. Sometimes, it is not possible to achieve separation with the chiral columns that are available, in which case the best approach is usually to derivatize the sample to form

TABLE 7.2   Three Approaches to the HPLC Separation of Enantiomers

Approach	Characteristics
Form diastereomers and separate by HPLC	Most likely to be successful, but most inconvenient in routine application; assays generally less precise; racemization possible
Add chiral complexing reagent to HPLC mobile phase	Least-used approach; chiral reagent is often expensive, adding to cost of assay; limited variety of suitable reagents available; reagent may interfere with detection
Use column with chiral reagent incorporated into stationary phase ("chiral column")	Most popular approach because of convenience and wide range of different chiral columns available; no practical problems, except achieving separation (usually through trial-and-error)

diastereomers. For a more detailed review of these various approaches to chiral separation, see Refs. (8–10a). In the remainder of this section we will summarize the more important aspects of method development for enantiomeric resolution.

Chiral Columns

Columns with chiral stationary phases are available from many suppliers (9) in about three dozen different forms. This range of availability is important, because the separation of a particular enantiomeric isomer pair depends almost totally on the chemistry of the sample/chiral reagent interaction. Changes in the mobile phase generally have only a minor effect on α for such separations; a change in stationary phase is usually the only way that separation can be much improved (after k'-values have been adjusted by varying solvent strength).

The main question in method development with chiral columns is: what type of column is most likely to separate a given sample? Because of the complexity of the sample/chiral reagent interaction, only rough guidelines can be given in most cases. It is useful to divide the various chiral columns into three categories, as summarized in Table 7.3.

Three-point interaction columns (see Fig. 7.1b) require two attractive interactions (usually hydrogen bonds) plus a third interaction that can be either attractive or repulsive. Chiral columns of this type are usually based on a sta-

TABLE 7.3   Classification of Chiral Columns

Column Type	Examples[a]	Sample Requirement
Three-point interaction	DNB-phenylglycine, DNB-leucine[b]	Two hydrogen-bonding groups in sample molecule (close to chiral center); aromatic or other bulky substituent group(s) in close proximity
Protein	BSA, α-1-acidglycoprotein	One or more polar or ionic groups close to chiral center
Cavity	Cyclodextrin, cellulose	Cyclic group close to chiral center that fits into cavity

[a]For sources of chiral columns, see Ref. (9).
[b]DNB refers to 2,4-dinitrobenzoyl.

tionary phase with an amide structure, as illustrated in Fig. 7.2a for a DNB-phenylglycine column. There are two carbonyls and two secondary amine groups in this structure, providing many possibilities for hydrogen bonding with the sample molecule. The sample molecule need only have two hydrogen-bonding groups (either donor or acceptor, or one of each), close to the chiral center, plus an aromatic ring. In some cases, a single hydroxyl group is sufficient (e.g., Ref. 11). The separation of an enantiomeric pair (Fig. 7.2b) on a DNB-phenylglycine column is shown in Fig. 7.2c.

If the sample molecule has the appropriate structural requirements for separation by three-point interaction (Table 7.3), but is unresolved on the first chiral column tried, it is usually worthwhile to try a different chiral column. Several studies (e.g., Refs. 13, 14) have shown that chiral columns of similar structure can exhibit large differences in chiral selectivity for a given sample. A good example is summarized in Fig. 7.3. Here several positional isomers of dihydrodimethylanthracene-diol were separated on four chiral columns (ionic or covalent forms of DNB-phenylglycine or -leucine). Here, the best column is the ionic phenyl glycine (3,4- and 5,6-isomers), or the ionic leucine (8,9- and 10,11-isomers). However, unsuccessful separations are noted for each column, depending on the sample. In another study (14), 23 amide samples were run on three different three-point interaction columns: 10 compounds were separated on one column, nine on another, and eight on the third. However, all but one of these compounds was separated ($\alpha > 1.05$) on one of the three columns. The conclusion is clear: when attempting chiral separations of this type, it is best to keep several different columns on hand for trial-and-error optimization of the separation.

Three-point interaction columns are generally the best chiral columns to try initially, if the sample molecule has the required structural features.

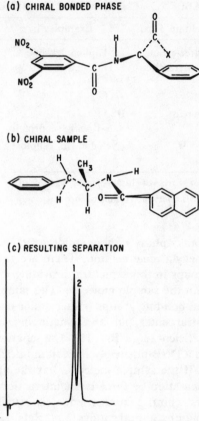

Figure 7.2   Chiral Separation on a DNB-Phenylglycine Column. Reprinted with permission from Ref. (12).

These columns were the first to be introduced and are, therefore, more fully developed (and, hopefully, more reliable). The use of other chiral column types (see below) often leads to wide and/or tailing bands that are less suited for routine quantitation, whereas bandwidths for three-point interaction columns are usually similar to those found with nonchiral HPLC columns (e.g., Figs. 7.2 and 7.3).

Protein columns (Table 7.3) use animal proteins as the chiral stationary phase. Carrier proteins such as albumin circulate in the bloodstream and are designed (or selected) by nature to recognize a variety of different enantiomers. The approach with proteins such as these is to bind them to a rigid matrix, which is then used as an HPLC column. These columns have numerous different binding sites that are specific for a wide range of compounds,

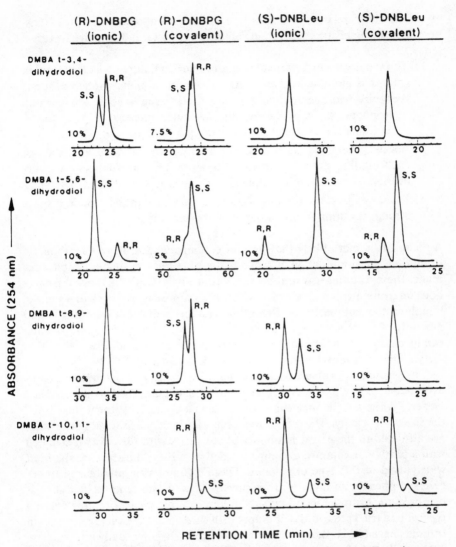

Figure 7.3 Separation of Various Dihydrodimethylanthracene-Diol Enantiomers, Using Different Chiral Columns. DNBPG, dinitrobenzoylphenylglycine; DNBLeu, dinitrobenzoyl-leucine. Reprinted with permission from Ref. (11).

especially those related to naturally occurring compounds that are transported in the bloodstream. Protein columns have several distinctive features:

1. The separation of a particular enantiomeric mixture on a given protein column is more likely than with other chiral columns; protein columns recognize hydrogen-bonding groups, polar groups, ionic groups, and hydrophobic areas in the sample molecule, and also respond to the three-dimensional structure of the sample molecule.
2. Because there are many different binding sites, the concentration of each binding site is rather low; therefore, protein columns are easily overloaded (maximum sample size is only a few nanomoles or less).
3. In many cases, the binding of sample molecules to the protein is quite strong, resulting in broad bands that often tail.

Although it is more likely that protein columns will result in enantiomeric resolution, the combination of broad bands plus small available sample sizes makes these columns less suitable for routine analysis. Enantiomeric separations on protein columns are also somewhat more sensitive to changes in the mobile-phase composition. Retention can be reduced by adding small amounts of propanol to the mobile phase, or by varying pH. Band spacing can be changed to a limited degree by the use of various additives: dimethyloctylamine, tetrabutylammonium ion, or octanoic acid (8).

Cavity columns (Table 7.3) achieve chiral selectivity by the presence of an asymmetric cavity in the chiral complexing agent attached to the surface. Generally, the sample molecule must possess a cyclic substituent that "fits" the size of the cavity. The requirement for additional substituent groups that are attracted to the chiral stationary phase varies with the cavity associated with a particular column. Columns bonded with β-cyclodextrin are the most widely used of this type of column. These columns can only be run under reversed-phase conditions, and are successful only with samples that possess one or more hydrogen-bonding substituent groups. Generally, this column type will be tried after initial attempts with three-point interaction or protein columns have failed. Bandwidths for cavity columns are usually intermediate between the latter column types, making cavity columns usable for routine analysis, but often less desirable than three-point interaction columns.

Derivatization to Diastereomers

Many chiral derivatizing reagents are available for reaction with different functional groups in the sample molecule (alcohols, amines, carboxylic acids, etc.). Successful resolution of the resulting diastereomers depends on (a) the proximity of the chiral reagent group to the chiral center of the sample molecule, and (b) HPLC separation conditions. Normal-phase HPLC is generally preferred for the separation of various isomers (positional, stereo, or diastereomeric), and resolution can be optimized as described in Sect. 4.5.

Selection of a suitable chiral derivatizing reagent is critical for a successful separation of enantiomers. However, the need for a manual derivatization step makes this procedure less attractive for routine analysis, because of the added labor and (in most cases) decrease in assay precision. The possibility of sample racemization during derivatization must also be considered. In addition, the relative rates of reaction between the derivatizing reagent and the sample compounds may be different [Ref. (15)], and this could affect the final analysis.

Mobile-Phase Additives

This approach is used less frequently than chemical derivatization. Cyclodextrin can be added to the mobile phase to achieve separations similar to those using cyclodextrin columns. Various metal-ligand complexing agents have been used to separate D- and L-enantiomers of amino acids (e.g., Fig. 5.14c). The use of 10-camphor-sulfonic acid as additive allows the enantiomeric resolution of various derivatives of 2-ethanolamine [Ref. (16)].

7.4   TRACE ANALYSIS

Somewhat different criteria generally are used in the development of analytical HPLC methods for components in the trace concentration range (< 100 ppm). The detailed aspects of trace analysis by HPLC have been given in Ref. (17). Trace analyses normally are conducted for a single (or a few) components in a relatively complex mixture. An example is the analysis of a pesticide residue in an extract of a plant or animal tissue, where the rest of the compounds in the sample are of no interest. The main requirement is to gain a clear separation of the trace component(s) (with good peak shape) from all possible interfering materials present in the sample, so that an accurate measurement of the desired constituent can be made.

General

In trace analysis, accuracy is the primary goal, rather than precision of the measurement. Trace analysis generally should be based on peak-height measurements, since this method of quantitation permits higher accuracy; that is, greater freedom from possible interference by neighboring constituents, compared to peak-area measurements. Less resolution is required for peak-height measurements for the same level of accuracy. Accuracy and reproducibility are also improved by using separation conditions that produce symmetrical (nontailing) peaks. Section 3.3 discusses the problems and remedies associated with tailing peaks.

Trace analysis usually is carried out using an isocratic separation, rather than gradient elution. Isocratic separations often allows higher sensitivity because of superior detector baselines. In addition, the equipment is simpler and analyses are performed in a shorter time compared with gradient separations. When routine trace measurements are needed for two or more components with widely varying retention, automated step gradients (a series of isocratic mobile phases, each of increasing strength) have been found useful (18).

Accurate trace analysis involves the selective measurement of the component(s) of interest. This can be accomplished with a chromatographic system that separates the compound(s) to be measured from other components in the sample mixture. Alternately, accurate measurements often can be made using a selective or specific detector for the component of interest. For example, use of an electrochemical detector for measuring trace amounts of the drug hydralazine in plasma is illustrated in Fig. 7.4. In this case, an internal standard (IS) was used in the procedure for quantitation, and salicylaldehyde hydrazone derivatives were made before chromatography to permit quantitation of the relatively unstable hydralazine (H). Note in Fig. 7.4 that the peaks of interest are free from interferences in this complex matrix, because of the high selectivity of the detector for the components of interest. Selective detectors greatly decrease the work needed to develop the desired trace analysis; therefore, a highly selective detector for the compound(s) of interest is recommended as a starting point in method development. Discussions of selective HPLC detectors are given in Refs. (20, 21); variable-wavelength UV, fluorimetric, and electrochemical detectors are widely used in trace analyses.

The method chosen for the trace-component separation should allow for direct injection of the sample of interest, if at all possible, to avoid baseline disturbances and retention variations. For example, reversed-phase chromatography should be considered first for aqueous samples; normal-phase is desirable for samples dissolved in nonpolar organic solvents. Before a reversed-phase separation can be attempted, less polar organic solvents (e.g., in ether or chloroform extracts of tissues) first should be evaporated to dryness; the residue containing the trace component must then be redissolved in an appropriate aqueous-organic mixture. The evaporation and sample transfer steps require special care to eliminate possible loss of the trace constituent.

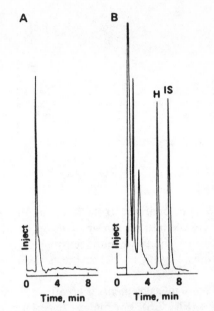

Figure 7.4   Trace Analysis of Hydralazine in Plasma with Electrochemical Detection (A) Extract of plasma control (1-mL sample); (B) Plasma control extract spiked with 100 ng each of hydralazine (H) and 4-methylhydralazine (internal standard IS) derivatives. Conditions: sample volume, 10 μL; column, Supelcosil LC-18-DB, 5-μm, 150 $\times$ 4.6 mm; mobile phase, 66% methanol in 0.055 M citric acid/0.02 M dibasic sodium phosphate (pH 2.5); flow rate, 1.5 mL/min; electrochemical detectors, Model 5100 A Coulochem; detector 1, $+0.25$ v; detector 2, $+0.60$ v; conditioning cell, $+0.20$ v; column temperature, 28° C. Reprinted with permission from Ref. (19).

Physical Effects

The sensitivity and accuracy of trace analysis often can be enhanced by proper selection of initial parameters. Trace separations should always be performed with highly efficient, small-particle columns, since the resulting sharp peaks provide the best separation of the peak(s) of interest from neighboring components. Columns with large plate numbers also enable higher sensitivity, since peak height h′ is proportional to the square root of column plate number: $h' \sim N^{1/2}$. Twenty-five-centimeter columns of 5-μm particles are preferred for trace analyses, since they provide a good compromise of N, analysis time, and operational convenience. Shorter columns (e.g., 15 cm) should be used when resolution is not limiting, since peak heights are inversely proportional to the square root of the column length.

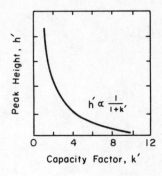

Figure 7.5   Effect of Capacity Factor k' on Peak
Height h'.

Columns should also be operated at the flow rate that produces the maximum plate number and the best resolution. This optimum flow rate also produces maximum peak height for a more sensitive trace measurement. For a 0.46-cm i.d. column of 5-μm particles, the desired flow rate typically is ~1.0 mL/min.

The sensitivity of trace measurements is greatly influenced by the k'-value of the peak of interest (Fig. 7.5). For highest sensitivity, the retention of the peak to be measured should be adjusted (usually by varying mobile-phase percent organic) so that it elutes in the range of 1 < k' < 3, if resolution from possible interferences permits. The relationship, h' ~ (1/1 + k'), predicts that significantly taller peaks will result at lower retention, but this must be balanced against generally poorer baselines and more frequent interferences as k' decreases.

Band-Spacing Effects

Even when column operating conditions for a separation have been optimized, overlap of the trace peak(s) of interest with other components in the mixture often occurs. In this case, separation selectivity must be altered to change the relative spacing of peaks (α-values) and allow the desired measurement to be made. The most powerful method for altering the spacing of peaks to eliminate peak overlap is to change mobile-phase strength and/or selectivity, as discussed in Chapters 2 and 4.

However, the process of optimizing mobile-phase selectivity in trace analysis sometimes can be simplified. In many cases the problem is to measure only a single constituent (or perhaps two or three) in a complex matrix. In such instances, if the initial separation is not adequate, band-spacing changes may be required only for the peaks adjacent to the peak(s) of interest. As is the case for all reversed-phase separations, the first step is to determine the appropriate strength of an acetonitrile/water mixture for the desired k′ elution value of the compound(s) of interest (see Sect. 4.3). (Acetonitrile is the preferred organic modifier because of its greater UV transmittance at low wavelengths, plus a lower viscosity that results in lower pressures and a higher plate number). If this optimum-strength acetonitrile/water mixture results in the overlapping of some other peak with the trace compound of interest, the acetonitrile concentration should be increased and decreased by 5%v to determine if band spacings are altered by solvent strength change. If the desired separation is not obtained with acetonitrile, an appropriate-strength methanol/water mixture should next be tested, followed by tetrahydrofuran/water, if the desired separation still is not obtained. Should these three solvent mixtures from the corners of the solvent-selectivity triangle fail to provide the needed separation of a single pair of components, then it is unlikely that any combination of the three organic modifiers will be successful. In this case, a different column stationary phase with an optimized mobile phase (as above) may provide the desired separation.

In reversed-phase chromatography, selectivity changes can often be achieved by using a different stationary phase, as discussed in Sect. 5.2. This is best accomplished by changing the type or chain length of the stationary phase, rather than the level of loading of the same phase on the support (i.e., partially- vs. fully-reacted support). In this case, changes in band spacings largely occur as a result of the difference in the strength of the stationary phase. Lower-strength stationary phases (e.g., cyano) require more water in the mobile phase for the same retention, and it is this change in water content in the mobile phase that is mainly responsible for the different spacing of peaks. Table 5.2 lists various reversed-phase stationary phases in order of their strength. The greatest change in selectivity or the spacing of peaks normally occurs between stationary phases with the greatest difference in their strength. Thus, if a C-18 column does not provide the desired reversed-phase separation, a cyano or phenyl column might show a selectivity change for the peaks of interest. In cases where the trace separation is restricted to a particular mobile phase (e.g., acetonitrile for UV detection at 205 nm, or a certain solvent needed for fluorimetric or electrochemical detection), a change in the stationary phase may be the only means by which significant band-spacing changes can be made.

Selectivity changes in reversed-phase also can be obtained with columns having the same organic stationary phase, but based on supports of different surface areas (or pore sizes). Thus, different band spacings can occur with a C-8 phase on a 350 m^2/g, 7.5-nm support, compared to the same C-8 phase on a 50 m^2/g, 30-nm support, because of the different level of water required in the mobile phase to obtain similar k′-values for the two packings.

As discussed in Sect. 5.2, changes in pH, ionic strength, and temperature can cause changes in band spacing, if the compound(s) of concern is ionic or ionizable. Unfortunately, selectivity changes of this type often are not predictable, since the peaks of interest may not be known. An effective experimental approach is to modestly increase and decrease the initial values of these variables to determine the level of their effect (e.g., vary temperature by $\pm 5°$ C; ionic strength $\pm$ fivefold; pH by $\pm$ one unit).

Finally, if all reasonable efforts fail to produce the needed separation, a different HPLC method may be required. It is not widely recognized that a greater range of selectivity differences is usually available in normal-phase chromatography (see Sect. 5.4) than for reversed-phase chromatography. This difference relates to the much greater range of chemical interactions that can occur in normal-phase separations. Normal-phase chromatography is particularly recommended if isomers are to be separated (22).

REFERENCES

1. D. T. Gjerde and J. S. Fritz, *Ion Chromatography*, 2nd ed., Verlag, Heidelberg, F.R.G., 1987.

2. R. E. Smith, *Ion Chromatography Applications*, CRC Press, Boca Raton, Fla., 1988.

3. W. W. Yau, J. J. Kirkland, and D. D. Bly, *Modern Size-Exclusion Liquid Chromatography*, Wiley, New York, 1979.

4. J. Janca, ed., *Steric Exclusion Liquid Chromatography of Polymers*, Marcel Dekker, New York, 1984.

5. T. Provder, ed., *Size Exclusion Chromatography*, ACS Symposium Series 245, American Chemical Society, Washington, D. C., 1984.

6. L. R. Snyder and J. J. Kirkland, *Introduction to Modern Liquid Chromatography*, 2nd ed., Wiley-Interscience, New York, 1979, Chapt. 10.

7. E. L. Johnson, R. Gloor, and R. E. Majors, *J. Chromatogr.*, *149* (1978) 571.

8. *J. Liquid Chromatogr.*, *Vol. 9 (2 and 3)*, (1986) (collection of papers on chiral HPLC).

9. R. Dappen, H. Arm, and V. Meyer, *J. Chromatogr.*, *373* (1986) 1.

10. W. Linder and C. Pettersson, in *Liquid Chromatography in Pharmaceutical Development: An Introduction*, I. M. Wainer, ed., Aster, Springfield, Ore., 1986, p. 63.

10a. M. Zief and L. J. Crane, *Chromatographic Chiral Separations*, Dekker, NY, 1988.

11. N. Oi and H. Kitahara, *J. Chromatogr.*, *265* (1983) 117.

12. I. W. Wainer and T. D. Doyle, *J. Chromatogr.*, *259* (1983) 465.

13. S. K. Yang and H. B. Weems, *Anal. Chem., 56* (1984) 2658.
14. N. Oi, M. Nagase, Y. Inda, and T. Doi, *J. Chromatogr., 265* (1983) 111.
15. I. S. Krull, in *Advances in Chromatography, Vol. 16,* Elsevier, New York, J. C. Giddings, E. Grushka, J. Cazes and P. R. Brown, eds., 1978, p. 176.
16. C. Pettersson and G. Schill, *J. Chromatogr., 204* (1981) 179.
17. L. R. Snyder, J. J. Kirkland, *Introduction to Modern Liquid Chromatography,* Wiley, New York, 1979, Chapt. 13.
18. F. Erni, R. W. Frei, and W. Lindner, *J. Chromatogr, 125* (1976) 265.
19. J. K. Wong, T. H. Joyce III, and D. H. Morrow, *J. Chromatogr., 385* (1987) 261.
20. R. P. W. Scott, *Liquid Chromatography Detectors,* 2nd ed., Elsevier, New York, 1986.
21. P. C. White, *Analyst, 109,* (1984) 677; 973.
22. L. R. Snyder and J. J. Kirkland, *Introduction to Modern Liquid Chromatography,* Wiley, New York, 1979, Chapt. 9.

8

COMPUTER-ASSISTED
METHOD
DEVELOPMENT

8.1 INTRODUCTION

In earlier chapters we have outlined certain strategies for method development in HPLC; each of these can be performed without the aid of a computer.
However, the increased emphasis on HPLC method development and optimization during the 1980s has been due, in large part, to our ability to automate
various aspects of the HPLC process. This computerized automation includes, (a) collecting and analyzing initial data, (b) organizing retention data
for further interpretation, (c) fitting these data to various models to predict
new experiments, and (d) finding the best conditions for a particular analysis
(1,2). In this chapter, we will describe those features of computer-aided
HPLC method development that are most useful for reaching these goals, as
well as specific computer programs that provide the best chance for success.

199

The use of computers for HPLC method development is not a substitute for good chromatography; software has no magic power to solve problems, and we are still a long way from "black box" method development. This fact is often overlooked by workers who consider the use of computer-assisted techniques for optimizing HPLC separations. What computers *will* do (if properly used) is to make some part of data collection, analysis, and/or final predictions more precise, faster, or easier to execute. For example, the calculation of retention times, peak widths, column plate number, and other parameters can all be done manually without an electronic data system. However, these manual calculations can require more time than is needed to make the experimental run. Similarly, optimization criteria such as relative resolution maps (Sect. 8.4) and retention mapping as a function of solvent composition can be carried out manually; however, the time required quickly discourages many users. However, while the computer is a tremendous help, effective method development must still be based on a reasonable strategy plus a skilled chromatographer.

There are currently three major uses for computers in HPLC method development: (a) instrument control, (b) data acquisition and analysis, and (c) prediction of separation as a function of experimental conditions (including selecting the "best" run). Most of the software for optimization and method development emphasizes the predictive aspects. An overview of some commercial software packages is described in Sect. 8.4.

8.2   INSTRUMENTAL CONTROL

The use of microprocessors for instrument control is now an important feature of modern HPLC systems. For example, microprocessors are used (a) in solvent-metering systems to blend solvents and provide precise flow control; (b) in detectors to maximize signal-to-noise ratios, and (if desired) select the wavelength (UV detector); (c) in ovens to control and change temperature; and (d) in autosamplers to reproducibly inject various samples onto the column at selected times. All of these features increase both the precision and accuracy of HPLC instruments, which are critical features for any method-development procedure.

Another useful aspect of current HPLC equipment is that many of these systems can be controlled from a central microcomputer. This capability allows the programming of specific conditions (e.g., gradients, temperature, wavelength selection), and also provides for automatically *changing* conditions from one sample to the next. This latter feature is quite useful for predetermined optimization strategies, where an entire set of experiments (varying conditions) can be designed to run automatically.

In addition, many computer-assisted method-development systems use a "learn as you go" approach, where the experimental conditions for the next separation depend on the results of previous runs. This type of system (e.g., sequential simplex design; see Sect. 8.4) requires a "learning" or interactive feedback control between instrument and controller, thus allowing the computer to determine and implement the next set of run conditions. A major benefit of any automated system is that it allows new separations to be carried out without further operator input. Therefore, the number of runs required for method development can be significantly increased without much effort. This capability greatly facilitates data gathering during the use of various computer-directed method-development procedures.

8.3 DATA ANALYSIS

A second important use of computers is for the analysis of raw data from the detector, to provide the final information needed for method development. Modern data systems supply software for determining bandwidth, peak height and peak area, and for simple calculations such as k', R_s, α, etc. Such capabilities often are essential, because method-development procedures require a measurement of separation quality for various peak pairs in the chromatogram. For software that uses a learning or iterative experimental approach, a single criterion is needed to characterize the quality of the separation. Several such criteria are detailed in Refs. (1,2).

8.4 PREDICTIVE SOFTWARE

Many procedures for computer-assisted HPLC method development have been described. These incorporate software that can make predictions and decisions for improving or optimizing a given separation. Detailed descriptions of many of these programs can be found in two recent books on optimization (1,2). In this section, we will review some of the major types of predictive software, with emphasis on programs that are commercially available.

Two main types of software are used in computer-assisted method development. Experimental-design systems assume some knowledge of the retention processes for various types of HPLC, and they use a predetermined experimental design to gather data. These data are then fit to some model, leading to predictions (mapping) for an optimum separation. Alternately, a learning approach can be used to carry out experimental runs. Based on the results of initial separations, the computer predicts appropriate experimental conditions for subsequent runs. These separations are carried out in turn, and the process is repeated until the desired separation is obtained.

Experimental-Design Systems I: DRYLAB®

Experimental-design systems are an extension of systematic method development, as described in Chapter 4. A first step in this approach is to choose an appropriate system (reversed-phase, normal-phase, ion-pair, etc.), and then establish the proper retention range for the sample. For isocratic separations, $1 < k' < 20$ is preferred; if there is a wider k' range, gradient elution is recommended (see Sect. 2.2 and Chapt. 6). In Sect. 4.2, a method is described for obtaining the best solvent strength: successive isocratic separations are performed, starting at 100% methanol and decreasing the methanol content by 20% for each subsequent injection. Although this technique is useful, it is tedious and inefficient. A computer program known as DRYLAB I (LC Resources, Lafayette, CA, run with IBM PC-compatible computers [3–6]), greatly simplifies this process for reversed-phase or ion-pair HPLC.

Method development with DRYLAB I begins with two gradient elution runs having different gradient times (all other conditions the same). Gradient-elution retention times and experimental conditions from these two runs are used by DRYLAB I to predict isocratic retention as a function of percent organic in the mobile phase. This procedure is based on the relationship:

$$\log k' = \log k_w - S\Phi, \tag{8.1}$$

where k_w is the extrapolated value of k' for the compound in pure water as the mobile phase, Φ is the volume fraction of organic in the mobile phase, and S is a constant that is characteristic of the organic solvent, the column, and the sample compound. DRYLAB I uses data from the two gradient runs to calculate k_w- and S-values for each compound in the sample. This information can then be used to predict values of Φ that position all peaks in the range of $1 < k' < 20$. (If the k'-values for components in the sample are outside this range, gradient elution must be used; see Fig. 9.6).

An example of this approach is shown in Fig. 8.1 for a mixture of nitro-compounds, using gradient elution with gradient times of 15 and 45 min (all other separation conditions constant). The experimental data shown in Table 8.1 permit the DRYLAB I software to calculate isocratic retention times for each of these compounds as a function of percent organic. This calculation, in turn, allows the predictions given in Table 8.2 for resolution, run time, and k' range as a function of mobile-phase composition. The data of Table 8.2 indicate that a reasonable k' range is provided by 50–70% methanol/water. Any mobile-phase composition in this range could be used for a final separa-

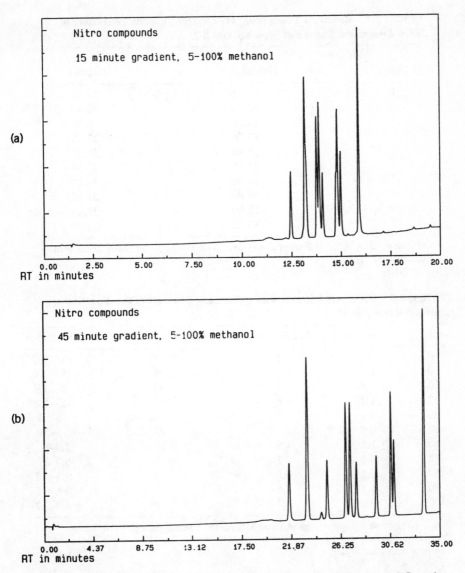

Figure 8.1 Two Initial Gradient Chromatograms for a Nitro-Compound Sample. Column, 25 × 0.46 cm, Zorbax C-8; mobile phase gradient, from 5% to 100% methanol/water; flow rate, 2.0 mL/min; temperature, 35° C. (a) Gradient time of 15 min; (b) Gradient time of 45 min. (See Fig. 8.2 for individual compounds.) Reprinted with permission from Ref. (3).

TABLE 8.1 Retention Times (min) for Gradient-Elution Separation of Nitro-Compound Sample of Figs. 8.1 and 8.2[a]

Peak	15-min Gradient	45-min Gradient
1	12.45	22.24
2	13.16	25.44
3	13.20	23.76
4	13.78	27.05
5	13.90	27.42
6	14.08	27.96
7	14.76	29.67
8	14.84	31.02
9	15.00	31.26
10	15.92	33.89

[a]Column, 25 × 0.46-cm Zorbax C-8 (5 μm); 5–100% methanol/water linear gradients; 2.0 mL/min.

TABLE 8.2 DRYLAB I Simulations of Isocratic Separation as a Function of Strong Solvent (%-B)[a]

%-B	R_s[b]	Band-Pair[c]	Run Time (min)	k′ Range
0	0.2	3,1	2973	118–2324
10	1.7	1,3	1159	57–906
20	1.8	4,5	453	27–353
30	1.2	9,8	177	13–138
40	0.1	9,8	70	6–54
50	0.8	2,3	28	3–21
60	1.1	4,5	12	1.4–8
70	0.0	2,1	5	0.7–3
80	0.2	3,8	3	0.3–1.2
90	0.1	6,8	2	0.1–0.5
100	0.0	3,7	1	0.0–0.2

[a]Nitro-Compound Sample of Figs. 8.1 and 8.2; calculated from data in Table 8.1.
[b]For N = 10,000.
[c]Worst-resolved band-pair.

tion, or as the starting point for further optimization. It should be noted that the identity of individual compounds in the sample need not be known when using DRYLAB I. When the total number of compounds in the sample is not known, optimization can proceed on the basis of the number of peaks observed; i.e., optimum conditions provide a maximum number of resolved bands in the final chromatogram.

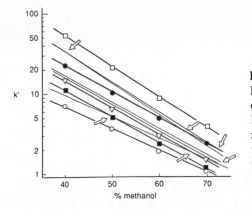

Figure 8.2   Plots of log k′ vs. Percent Methanol/water (v/v) for a Mixture of Nitro Compounds. Conditions as in Figure 8.1. Compounds: nitrobenzene, 2,6- dinitrotoluene, benzene, 2-nitrotoluene, 4-nitrotoluene, 3-nitrotoluene, toluene, 2-nitro-1,3-xylene, 4-nitro-1,3-xylene, 1,3-xylene. Reprinted with permission from Ref. (3).

An alternative permitted by DRYLAB I is to use data from two *isocratic* separations having different percent organic values. This approach gives the same predictions as in Table 8.2 and is a quantitative extension of the manual method described in Sect. 4.2. Whereas only two gradient experiments are sufficient to define sample retention as a function of percent organic, more than two isocratic separations may be required in order to find a suitable pair of runs giving reasonable retention (for input to DRYLAB I). The procedure of Tables 8.1 and 8.2 (based on two gradient runs) is therefore a more efficient approach for optimizing k′; however, it suffers from slightly reduced accuracy in predictions of k′ (see discussion in Refs. 7 and 8), compared with the use of two isocratic runs. However, this reduced accuracy is often not significant for practical method development.

DRYLAB I software has additional capabilities besides predicting an appropriate mobile phase strength (percent organic). As discussed in Sect. 4.2, changes in percent organic in reversed-phase separation can lead to significant band-spacing variations. This effect can be exploited using a technique called *Fast-Method Development* (3). For example, to define the retention characteristics of each component of the nitro-compound sample described above, the data of Table 8.2 can be plotted as shown in Fig. 8.2. Several peak crossovers occur as the percent organic is varied (arrows). The optimum mobile-phase composition can be determined with a relative resolution map (RRM) provided by the DRYLAB I software. An example is shown in Fig. 8.3; the RRM is calculated by the computer from the standard resolution equation (Eqn. 2.3), assuming a constant value of N (10,000 plates) for all compounds in the sample.

The RRM of Fig. 8.3 shows that for certain mobile phase mixtures (e.g., 50% methanol), no separation is observed for at least one band pair (B and C). However, mobile phases having between 53% and 60% methanol (%B)

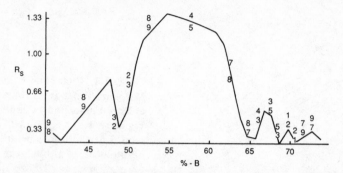

Figure 8.3   Relative Resolution Map (RRM) for Nitro Compound Sample of Fig. 8.1 (DRYLAB I simulation). Numbers on map indicate adjacent peaks with the limiting (lowest) resolution for each mobile-phase composition. Reprinted with permission from Ref. (3).

offer reasonable resolution ($R_s > 1.25$) for all peaks. A final optimized separation with 55% methanol is shown in Fig. 8.4. Figure 8.4a shows the separation predicted by DRYLAB I, and Figure 8.4b shows the experimental chromatogram for the same conditions.

The Fast-Method Development (DRYLAB) approach relies on band-spacing changes that occur as a result of varying isocratic solvent strength. The same technique can be used in gradient-elution method development (DRYLAB G software [9–11]). Again, two gradient runs are performed using different gradient times. In this case, the results are used to predict gradient-elution separation as a function of changes in gradient time, initial and final percent organic in the mobile phase, and gradient shape. An illustration of this procedure using the DRYLAB G software is shown in the following example.

A sample containing 16 polyaromatic hydrocarbons (PAH) was first run using two different gradient times (20 and 60 min) in an acetonitrile/water system. Figure 8.5 shows the chromatograms for these 5–100% acetonitrile gradients. All 16 compounds are partially resolved in the 20-min gradient, but only 15 peaks appear in the 60-min run (bands 14 and 15 co-elute). The retention times and experimental conditions from these two runs can be entered into DRYLAB G for the computer simulation of additional runs for method development.

Changes in gradient time are equivalent to variations in percent organic in an isocratic run, so far as retention (k') and selectivity (α) are concerned. Therefore, a change in gradient time can have a dramatic effect on the resolution of adjacent compounds. This can be seen in Fig. 8.6, where a relative resolution map (10,000-plate column) is shown for this PAH sample. Opti-

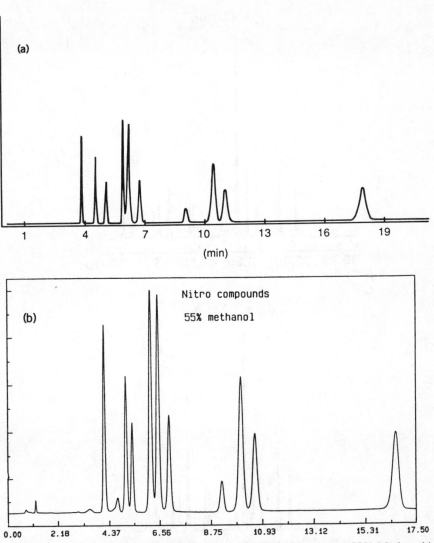

Figure 8.4 Isocratic Separation of Nitro Compound Sample Using a 55% Methanol/Water Mobile Phase. Other conditions as in Fig. 8.1. (a) DRYLAB I simulation; (b) experimental chromatogram. Reprinted with permission from Ref. (3).

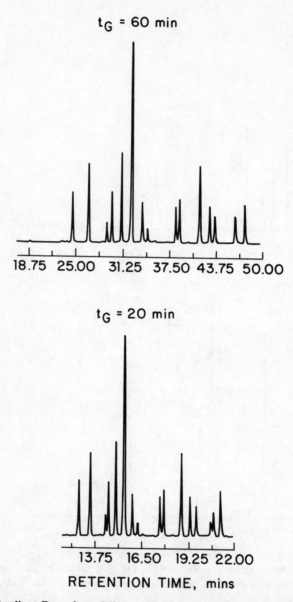

Figure 8.5 Gradient Runs for a Mixture of 16 Polyaromatic Hydrocarbons Using Different Gradient Times. Gradients, 5–100% acetonitrile/water, Column, 15 × 0.46 cm Supelcosil LC-PAH; 2 mL/min; 35° C. Reprinted with permission from Ref. (11).

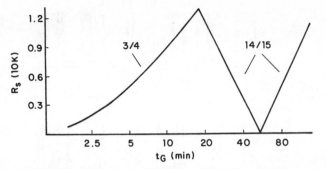

Figure 8.6   Relative Resolution Map (RRM) for PAH Sample of Fig. 8.5 (DRYLAB G). Predicted resolution assumes a column plate number of 10,000.

mum resolution is found for a gradient time of 20 min; shorter gradient times result in an increasing overlap of bands 3 and 4, and longer times result in greater overlap of bands 14 and 15.

At this point, a gradient time of 20 min could be chosen. However, using the DRYLAB G software, further options can be explored with the data already available (Fig. 8.5). To obtain more accurate predictions, an estimate of the actual plate number of this column is needed. This can be obtained by examining the separation of bands 14 and 15 in the 20-min run, which have $R_s = 1.03$, estimated by using Table 2.1. After taking this estimated resolution into account, DRYLAB G predicts that the plate count is 7,140 for the column in use. This value is then used in all subsequent calculations by the computer.

The next step is to optimize the gradient range; i.e., the percent organic at the start and end of the gradient. DRYLAB G predicts that a 44–94% acetonitrile gradient will be optimum (this allows for a 5% change in organic prior to the first band, and 5% after the last band). To maintain an optimum gradient slope (recall: 5–100%, or 95% acetonitrile in 20 min for the original gradient), a new gradient time is calculated by DRYLAB G: [(94–44)/(100–5)] × 20 = 10 min for the 44–94% acetonitrile run. The minimum resolution predicted for this separation is $R_s = 1.11$ for bands 14 and 15; the simulated chromatogram for this case is shown in Fig. 8.7a.

Further improvements in resolution can be obtained by fine-tuning the gradient range (mainly by changing the starting %-organic), or by using nonlinear gradients (segmented gradients). A series of trial-and-error simulations suggested a 44–60% linear gradient in acetonitrile in 7 min, followed by a 60–100% linear gradient in 2 min, for a total gradient time of 9 min. This

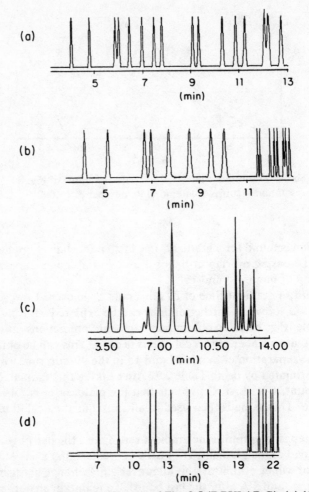

Figure 8.7 Chromatograms for Sample of Fig. 8.5 (DRYLAB G). (a) 10-min gradient, 44-94% acetonitrile/water; (b) 7-min gradient, 44-60% acetonitrile/water followed by a 2-min gradient from 60-100% acetonitrile/water; (c) actual separation using conditions of (b); (d) same simulation as in (b) except a 25 × 0.46-cm column was assumed. Other conditions as in Fig. 8.5.

separation was predicted to have a minimum $R_s = 1.26$, as shown in Fig. 8.7b. The actual separation in Fig. 8.7c shows good agreement with this simulation. The degree of additional improvement can be predicted by DRYLAB G for a 25-cm column in place of the original 15-cm column (Fig. 8.7d).

The DRYLAB software is a valuable tool for optimizing the mobile phase for either gradient or isocratic runs. These same programs also allow the further improvement of separation, by varying column conditions: column dimensions, flow rate, and particle size (see "Predictions of Plate Number N" in this section). This combined approach may produce the desired separation directly, or it can be used as a first step in further method development involving other variables.

Fast-Method Development (based on DRYLAB software) has proven useful in many systems where the final separation is not very difficult; i.e., "easy" or "average" samples, as in Fig. 4.5. This is often the case when there are fewer than 10 compounds in the sample. For more difficult cases, one of the following computer-aided systems may be necessary.

Experimental-Design Systems II: Mixture-Design Statistical Technique (MDST)

Computer-aided method development based on solvent-selectivity optimization was first described in 1980 (12). This approach, referred to as a *mixture-design statistical technique* (MDST), was based on the solvent-selectivity-triangle approach (13) described in Sect. 2.3. The MDST method assumes that maximum changes in band spacing will result from the use of three different organic solvents in the mobile phase, in turn leading to the best separation. Initially, using reversed-phase HPLC, it was assumed that an appropriate k' range could be obtained by varying organic composition (solvent strength) in steps, as described in Sect. 4.2. However, the use of DRYLAB I is currently preferred for selecting an appropriate solvent strength (using two gradient runs).

Chromatograms are next obtained with seven mobile phases comprising mixtures of three organic solvents (methanol, acetonitrile, and tetrahydrofuran) plus water. Retention times t_R are determined for each compound of interest in all seven solvent mixtures. The retention of each compound is then mapped over the entire solvent-composition range (constant solvent strength), and the optimum separation is predicted for some intermediate mobile-phase composition. The actual procedure is illustrated qualitatively in Sect. 4.3 for three examples: steroids, herbicides, and photochemicals.

The MDST method does not require a computer; manual inspection of the chromatograms can be used to develop an adequate separation. However, a more convenient and precise picture of retention behavior can be obtained using recently published software (14); this software has since been modified and made commercially available by LC Resources Inc., Lafayette, Calif. (DRYLAB S). The experimental retention for each compound in the sample is fitted to a model (usually a quadratic equation) as a function of solvent composition. An example of this fit is shown in Fig. 8.8 for 2-methoxynaphthalene on a C-8 column; retention time is plotted as a function of mobile-phase composition. When retention (k' or t_R) for all compounds can be predicted accurately (this is the case for all reversed-phase systems investigated to date), it is straightforward to examine all solvent combinations and select the one that provides the best separation (using any desired separation criterion; R_s, α, etc.)

An example of this approach for separating a mixture of nine substituted naphthalenes is shown in Fig. 8.9. To obtain this result, retention time and bandwidth data (or an estimate of column plate number) were entered into the MDST program. The retention for each peak and resolution between all peak pairs was then mapped for all mobile-phase conditions contained within the confines of the solvent triangle. In this case, it was assumed that all compounds in the sample needed to be resolved in the final method. Figure 8.10 shows resolution maps for all eight peak pairs; shaded regions are areas of solvent composition where $R_s < 1.5$ for that particular peak pair. Figure 8.11 is an overlay of these eight resolution maps, called an *overlapping resolution map* (ORM). It reveals one region (in white) where all peaks have at least $R_s > 1.5$. The software also predicts the optimum mobile-phase composition (minimum R_s-value for all compounds; shown here as an "X" at the bottom of the triangle). The experimental chromatogram in Fig. 8.12 confirms this prediction.

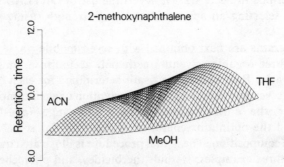

Figure 8.8   Retention Time Plot for 2-Methoxynaphthalene as a Function of Mobile-Phase Composition. Zorbax C-8 column, 15 × 0.46 cm; flow rate, 2.0 mL/min; temperature, 40° C. Data from Ref. (12).

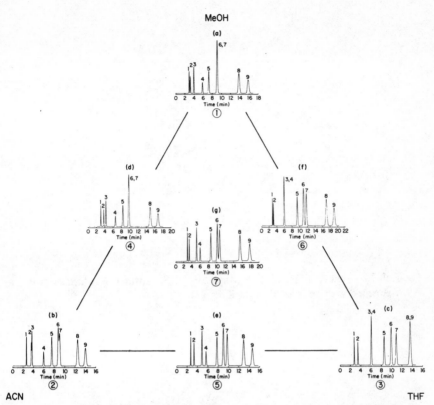

Figure 8.9 Separation of Nine Substituted Naphthalenes with Seven Mobile-Phase Compositions for the MDST Method. Conditions as in Fig. 8.8. Reprinted with permission from Ref. (12).

In the MDST approach, resolution values generally are used as the criteria for separation. However, it also is possible to calculate α values or retention-time difference (Δt_R) as a basis for separation. In addition to its use for reversed-phase HPLC, the current software has been used for ion-pair (15) and normal-phase (16–18) HPLC, as well as for gradient-elution optimization (19–20) and combined solvent/column optimization (21).

As described in Chapter 4, the use of solvent selectivity for method development and separation optimization is well established. The MDST software requires at least seven experiments (for a good quadratic fit of the data), plus additional runs to identify or confirm peaks as necessary. The number of additional runs varies with sample complexity. However, individual compounds need not be run separately. A smaller number of submixtures may serve to

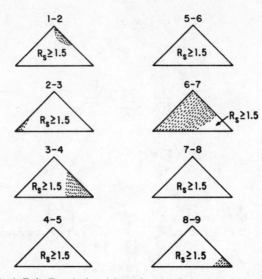

Figure 8.10 Peak-Pair Resolution Maps for All Substituted Naphthalenes in Fig. 8.9. Reprinted with permission from Ref. (12).

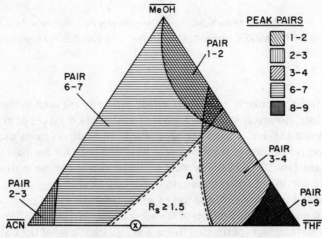

Figure 8.11 Overlapping Resolution Map Based on Fig. 8.10. Dashed lines (---) are estimate of resolution mapping precision. Reprinted with permission from Ref. (12).

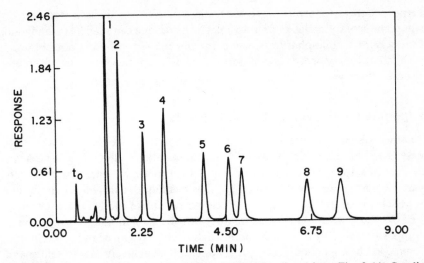

Figure 8.12 Optimum Separation Based on ORM Predicted from Fig. 8.11. Conditions as for Fig. 8.8 except mobile phase, 32/15/53 acetonitrile/tetrahydrofuran/water. Reprinted with permission from Ref (12).

identify the positions of each compound in each run. Alternatively, the individual components can be identified by other means, for example, by on-line spectroscopic examination using a diode-array detector. Still another approach is to formulate two sample mixtures having different concentrations of each solute in the two samples, and run each sample. Peak-height (or peak area) ratios for a given compound in each run are then predictable (equal to the concentration ratio in each sample). If the components in the mixture are unknown, detection at two wavelengths (e.g., 260 and 230 nm) often can be useful in identifying peaks.

Once the initial retention data are obtained, computer simulations can be performed to determine the separation for other conditions within the boundaries of the original three mobile phases (corners of the triangle). Alternatively, if one solvent is not desired in the final method (e.g., THF for detection using low-wavelength UV), it can be excluded or minimized, and (only) other areas of the solvent triangle investigated for the best separation.

Another approach to this type of optimization has been proposed by de Galan and co-workers (22–24) for reversed-phase HPLC. Their approach is to first run the binary mixtures of each organic solvent (methanol, acetonitrile, and tetrahydrofuran) plus water, and to then proceed to the ternary mixtures in a systematic manner. The data analysis initially assumes a linear relationship between retention and solvent composition; however, further experiments are then performed in regions where nonlinearity is indicated and/or where the separation looks promising. The technique can then be extended to four-solvent mobile-phase systems if the desired separation is not obtained with either binary or ternary mixtures. A similar approach has also been applied by these workers for ion-pair chromatography (25–26).

Experimental-Design Systems III: PESOS

Another system that incorporates a predesigned set of experiments is marketed as PESOS (Perkin-Elmer Solvent Optimization System [27]). The strategy for this approach is shown in Fig. 8.13; experiments are systematically carried out to map solvent selectivity. This technique is also based on the solvent-selectivity triangle (Fig. 2.16). However, unlike the MDST approach, PESOS starts with a particular mixture of solvents and modifies the solvent composition slightly with each run until the "best" (or an adequate) system is found. With proper equipment, this approach is relatively easy to set up and automate. Any criterion for separation (resolution, optimization functions, etc.) can be used to guide the development, since many experiments are performed before final data analysis.

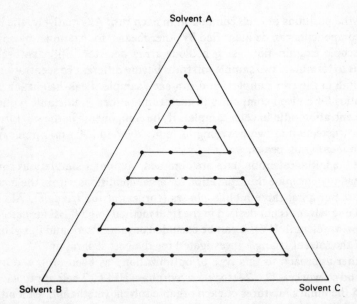

Figure 8.13 PESOS, Experimental Design of Perkin-Elmer Solvent Optimization System. Reprinted with permission from Perkin-Elmer Corp.

Unfortunately, few examples of this technique have been reported in the scientific literature. This approach is essentially a "brute force" procedure, where retention mapping is carried out systematically by changing the mobile-phase composition in small, fixed increments. This gridding technique is less efficient than the MDST procedure of Figs. 8.8–8.10; i.e., it usually requires many more experimental runs.

Learning Systems

Another class of computer-aided strategies involves those in which experiments are performed and evaluated immediately after they are run. Based on the results of the most recent run(s), additional experimental conditions are defined and performed, until an adequate or optimum separation is obtained. The most widely used systems of this type are simplex designs, reported initially by Smits and co-workers (28) and later by many other groups. Simplex approaches for HPLC method development have been reviewed in two recent books (1,2).

Simplex designs assume no model for retention; therefore, they are not limited to any set of variables. Indeed, they are excellent choices for analytical systems where it is not initially clear which variables, and what levels of each variable, are likely to be important in changing the response to the system. As such, simplex can be quite useful for determining a useful area on which to focus more detailed method development, if such information is desired. However, we believe that those variables most likely to be useful in HPLC method development generally are known or can be easily predicted. Therefore, predesigned experiments that use known information about the system (described in sections above) are more useful and efficient. Nevertheless, simplex-design experiments are available in several commercial HPLC systems such as OPTIM I (Spectra Physics), SUMMIT (Bruker), and TAMED (LDC, Milton Roy).

Next, we will describe an example of simplex method development (29) based on a precursor to the OPTIM I and TAMED systems. The reversed-phase separation of four substituted phenols (before optimization) is shown in Fig. 8.14 for a methanol/water (acetic acid) mobile phase. This binary-solvent system was expanded to a ternary-solvent-system search for a better separation of these four compounds. A suitable chromatographic resolution function (CRF) was used as a criterion to evaluate the results of each separation. In addition, it was necessary to place boundary conditions on some of the variables (e.g., percentage of all solvents must equal 100%, and no solvent can be <0%, conditions that can easily occur if the simplex is calculated by the computer).

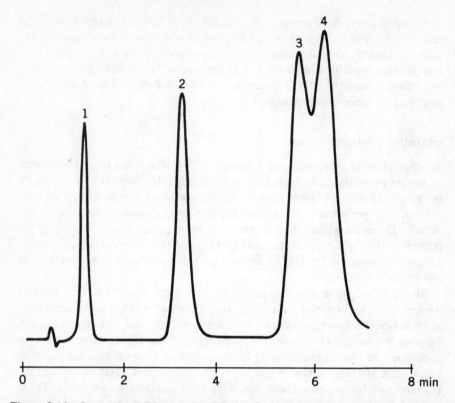

Figure 8.14 Separation of Substituted Phenols before Simplex Optimization. Mobile phase, methanol/water/acetic acid, 55/45/0.1; compounds: (1) methyl p-hydroxybenzoate, (2) propyl p-hydroxybenzoate, (3) butyl p-hydroxybenzoate, (4) propyl gallate. Reprinted with permission from Ref. (29).

The progress of simplex optimization for this sample is shown in Fig. 8.15, where the CRF is plotted as a function of the experiment number. A computer program TERNOPT is used here to plan the successive experiments in the simplex design. The mobile-phase compositions of the experiments are shown in Fig. 8.16, starting from the lower right corner of the figure (90% water, 1% methanol, 9% acetonitrile), and proceeding eventually to an optimum of 52% water, 21% methanol, and 27% acetonitrile in 30 experiments. The final separation is shown in Figure 8.17 for the four compounds, with the last peak eluting at 6.1 min.

The example in Fig. 8.17 shows that a reasonable mobile phase was gener-

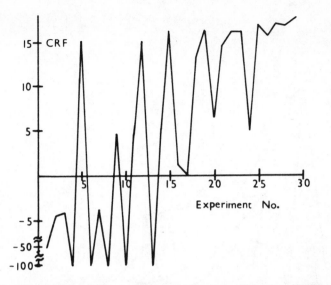

Figure 8.15 Relationship of Chromatographic Response Function (CRF) with Experiment Number during Optimization with TERNOPT Program. Reprinted with permission from Ref. (29).

ated for this sample. However, this process also demonstrates both the advantages and disadvantages of simplex optimization for HPLC. An advantage is that the entire procedure can proceed automatically, provided that the HPLC instrument is directly connected to the computer predicting each subsequent experiment. Boundary conditions must be described at the beginning of the procedure, and some measure of the quality of each run (such as a CRF) must be used to decide how to proceed with the next experiments. If these constraints are properly established, the optimization procedure can proceed unattended without prior knowledge of the system.

A major disadvantage of simplex optimization is that it often requires many experiments. In the example just described (Fig. 8.15), 30 runs were needed to move from a binary, isocratic separation to the final ternary system with only a modest increase in the quality of the overall separation. Another disadvantage of simplex systems is that total instrument control is needed, because of the necessary feedback from one experiment to the next. Therefore, the user must have a complete system from a single vendor to use such programs.

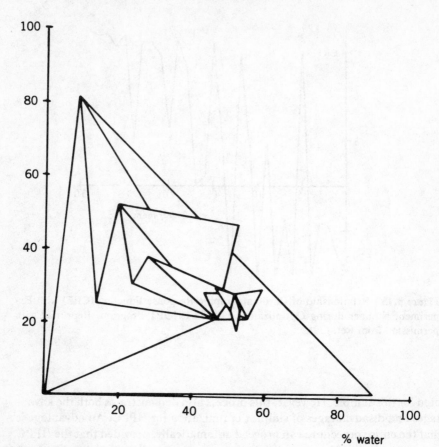

% water

Figure 8.16 Simplex Optimization with TERNOPT Program. Reprinted with permission from (29).

Predictions of Plate Number N: DRYLAB I and G

The computer-aided method-development procedures discussed above are based on varying mobile-phase or gradient conditions, in turn leading to the optimization of retention (k′ and α). Further improvements in separation (Eqn. 2.3) must therefore involve kinetic optimization; i.e., changes in the plate number N. In some cases it may be necessary to further increase resolution R_s; in other cases it may be desired to reduce run time, increase detection sensitivity (by decreasing bandwidth), or reduce the column pressure.

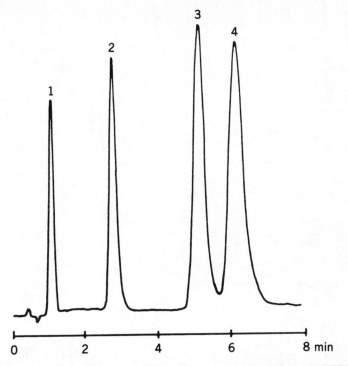

Figure 8.17 Final Separation of Substituted Phenols Optimized with TERNOPT Program. Mobile phase, 21/27/52/0.1 methanol/acetonitrile/water/acetic acid; compounds as in Fig. 8.14. Reprinted with permission from Ref. (29).

TABLE 8.3 Column Simulation Results for Mixture of Four Steroids[a]

Column Type	Flow Rate (mL/min)	R_s	Run Time (min)	Pressure (psi)	Sensitivity (mL^{-1})[b]
15-cm, 6-μm	1.75	3.0	10.9	969	5.0
5-cm, 6-μm	1.75	1.7	3.6	323	8.5
4-cm, 3-μm	5.0	2.3	1.8	1625	7.8

[a]Summary of DRYLAB I optimization of column conditions for steroid sample of Fig. 8.18.
[b]"Sensitivity" refers to reciprocal peak volume.

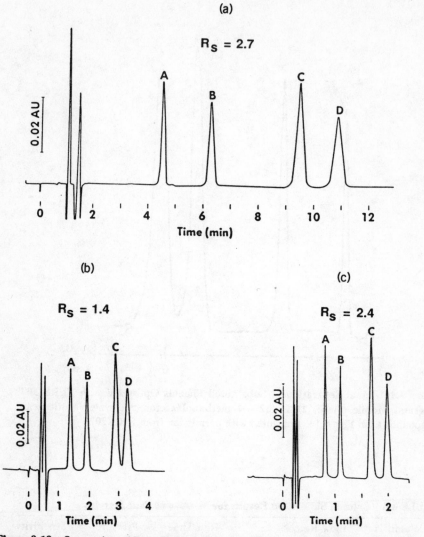

Figure 8.18 Separation of Four Steroids Using Different Column Configurations After Reversed-Phase Retention Optimization. (a) 15 × 0.46-cm column, 6-μm particles; (b) 5 × 0.46-cm column, 6-μm particles; (c) 4 × 0.62-cm column, 3-μm particles; mobile phase, 24/10 (v/v) methanol/tetrahydrofuran in water; flow rate, (a) and (b), 1.75 mL/min, (c) 5.0 mL/min; detection, 254 nm; 50° C; compounds, (A) prednisone, (B) hydrocortisone, (C) corticosterone, (C) cortexolone. Reprinted with permission from Ref. (31).

The DRYLAB programs provide for both retention optimization and the simulation of changes in column conditions (3-11,30). Retention optimization is normally carried out first, using any of the techniques described in this chapter. Then DRYLAB is used to carry out further simulations, where relative retention (k' and α) are held fixed, and column conditions (and N) are varied. The utility of this approach is illustrated by the following example.

An initial separation of four steroids was obtained on a 15×0.46-cm column of 6-μm Zorbax C-8 packing, as shown in Fig. 8.18a. Solvent strength (percent organic) has been optimized at this point (using MDST), but the separation time is longer than necessary; this separation is "too good." DRY-LAB I was used next to explore the effect of changes in column length, flow rate, and particle size on the separation. Table 8.3 shows a summary for three different sets of DRYLAB simulations (the initial run and two simulated experiments). Acceptable resolution ($R_s > 1.5$) is observed for shorter columns and/or smaller particles—with a reduction in separation time by a factor of three- to fivefold. The experimental chromatograms in Figs. 8.18b,c show good agreement with the predictions of Table 8.3 (compare R_s-values in Table 8.3 with those in Fig. 8.18).

Summary

The computer-optimization or method-development procedures described here are listed in Table 8.4. The features, strengths, and weaknesses of each approach are summarized for application to a given separation problem. Our present belief is that method development should begin with solvent-strength optimization, as discussed in Chapter 4 (see Fig. 4.5). This suggests the use of software such as DRYLAB (for either isocratic or gradient procedures). If this approach is unsuccessful in obtaining adequate separation, then the mobile-phase components should be varied. Optimization of this step can be carried out (6) either with the DRYLAB software using discrete changes in the mobile phase (e.g., substitute methanol for acetonitrile, change pH, etc.), or use can be made of the four-solvent MDST procedure.

When sample retention is adequate for a successful separation, further fine-tuning of column conditions can be carried out using software such as DRYLAB. Other approaches as noted in Table 8.4 can also be tried.

TABLE 8.4 Summary of Features, Strengths, and Limitations of Various Computer-Optimization Approaches to HPLC Method Development[a]

Procedure and Comments

DRYLAB (SOLVENT-STRENGTH OPTIMIZATION)

1. Carries out solvent-strength optimization using computer simulations (semi-manual search for best conditions). Can be repeated for different organic solvents, pH, etc.
2. Requires only two experimental runs; relatively easy to use; provides acceptable separation for the majority of "easy" or "average" samples; commercial software runs on a PC
3. Solvent-strength optimization is less powerful than four-solvent mapping; some samples not separable in this way

FOUR-SOLVENT MAPPING (MDST)

1. Carries out mobile-phase optimization by using an efficient experimental-design procedure; can vary types of solvents, pH, etc.; can be run in fully automated mode
2. Very powerful and practical means of maximizing resolution for a given sample; this software published, widely applied, and thoroughly tested
3. Normally requires a large number of experimental runs to carry out; e.g., 20–50 injections of sample and standards; can be inaccurate for separations involving pH variation

FOUR-SOLVENT MAPPING (PESOS)

1. Carries out mobile-phase optimization by systematic mapping ("gridding") of retention vs. mobile-phase composition; same variables possible as for MDST
2. Powerful and straightforward procedure for maximizing sample resolution; similar capabilities as MDST; simple approach, so little can go wrong
3. Inefficient use of experimental data, therefore requires many experimental runs for each sample; computer program requires Perkin-Elmer HPLC System

LEARNING SYSTEMS (OPTIM I, SUMMIT, TAMED)

1. Carries out mobile-phase optimization by systematic trial-and-error approach; some versions also allow changes in other conditions (temperature, flow rate, column dimensions, etc.)
2. Few advantages compared with other approaches, except possibly the flexible choice of parameters for simulataneous optimization
3. Very inefficient at optimization; requires many runs to obtain modest improvement in separation; only practical when incorporated in specific hardware that carries out all runs automatically

DRYLAB (COLUMN OPTIMIZATION)

1. Can be used after mobile-phase optimization using any of the above procedures; predicts change in separation that results from change in column dimensions, flow rate or particle size
2. Requires only one experimental run to carry out predictions; relatively easy to use
3. Predictions of separation can be in error, due to nonideal behavior of sample (provisions for correcting such errors incorporated into DRYLAB)

---

[a]Applicable for both isocratic and gradient methods.

REFERENCES

1. J. C. Berridge, *Techniques for the Automated Optimization of HPLC Separations*, Wiley-Interscience, New York, 1985.
2. P. J. Schoenmakers, *Optimization of Chromatographic Selectivity*, Elsevier, Amsterdam, 1986.
3. L. R. Snyder, J. W. Dolan, and M. P. Rigney, *LC-GC, 4* (1986) 921.
4. M. A. Quarry, R. L. Grob, L. R. Snyder, J. W. Dolan, and M. P. Rigney, *J. Chromatogr., 384* (1987) 163.
5. L. R. Snyder, J. W. Dolan, and M. A. Quarry, *TrAC Trends Anal. Chem. (Pers. Ed.), 6* (1987) 106.
6. L. R. Snyder, M. A. Quarry, and J. L. Glajch, *Chromatographia, 24* (1987) 33.
7. M. A. Quarry, R. L. Grob, and L. R. Snyder, *Anal. Chem., 58 (1986)* 907.
8. L. R. Snyder and M. A. Quarry, *J. Liq. Chromatogr., 10* (1987) 1789.
9. J. W. Dolan and L. R. Snyder, *Chromatogr. Mag., 2* (1987) 49.
10. J. W. Dolan and L. R. Snyder, *LC-GC, 5* (1987) 970.
11. J. W. Dolan, L. R. Snyder, and M. A. Quarry, *Chromatographia, 24* (1987) 261.
12. J. L. Glajch, J. J. Kirkland, K. M. Squire, and J. M. Minor, *J. Chromatogr., 199* (1980) 57.
13. L. R. Snyder, *J. Chromatogr. Sci., 16* (1978) 223.
14. J. L. Glajch, J. J. Kirkland, and J. M. Minor, *J. Liquid Chromatogr., 10* (1987) 1727.
15. A. P. Goldberg, E. L. Nowakowska, P. E. Antle, and L. R. Snyder, *J. Chromatogr., 316* (1984) 241.
16. L. R. Snyder, J. L. Glajch, and J. J. Kirkland, *J. Chromatogr., 218* (1981) 299.
17. J. L. Glajch, J. J. Kirkland, and L. R. Snyder, *J. Chromatogr., 238* (1982) 269.
18. P. E. Antle, *Chromatographia, 15* (1982) 277.
19. J. L. Glajch and J. J. Kirkland, *Anal. Chem., 54* (1982) 2593.
20. J. J. Kirkland and J. L. Glajch, *J. Chromatogr., 255* (1983) 27.
21. J. L. Glajch, J. C. Gluckman, J. G. Charikofsky, J. M. Minor, and J. J. Kirkland, *J. Chromatogr., 318* (1985) 25.
22. P. J. Schoenmakers, A. C. J. H. Drouen, H. A. H. Billiet, and L. de Galan, *Chromatographia, 15* (1982) 688.
23. A. C. J. H. Drouen, H. A. H. Billiet, P. J. Schoenmakers, and L. de Galan, *Chromatographia, 16* (1982) 48.
24. A. C. J. H. Drouen, H. A. H. Billiet, and L. de Galan, *J. Chromatogr., 352* (1986) 127.
25. A. Bartha, Gy. Vigh, H. A. H. Billiet, and L. de Galan, *Chromatographia, 20* (1985) 587.
26. H. A. H. Billiet, J. Vuik, J. K. Strasters, and L. de Galan, *J. Chromatogr., 384* (1987) 153.

27. M. W. Dong, R. D. Conlon and A. F. Poile, *Amer. Lab. Mag.*, May, 1988, p. 48.

28. R. Smits, C. Vanroelen, and D. L. Massart, *Z. Anal. Chem.*, *273* (1975) 1.

29. J. C. Berridge, *J. Chromatogr.*, *244* (1982) 1.

30. L. R. Snyder and J. W. Dolan, *Am. Lab. (Fairfield Conn.)*, August (1986) 37.

31. L. R. Snyder and P. E. Antle, *LC Mag.*, *3* (1985) 98.

9

METHOD DEVELOPMENT PROCEDURES

9.1 **Introduction**

9.2 **Isocratic Reversed-Phase Separations**
   General Conditions for All Reversed-Phase Separations

9.3 **Reversed-Phase Separations by Gradient Elution**

9.4 **Isocratic Normal-Phase Separations**

9.5 **Ion-Pair Separations**

9.1 INTRODUCTION

In this chapter we will summarize in "recipe" form the specific steps that should be taken to develop a particular separation method. The first procedure (for reversed-phase separations, Sect. 9.2) assumes (a) that a relatively simple mixture of small molecules (i.e., ≤ 10 components) is involved, (b) that gradient elution and a computer are not available, and (c) that the separations are to be carried out and analyzed manually. To illustrate the steps that are used in this method-development procedure, flowsheets are given in Tables 9.1–9.3.

In Sect. 9.3 we summarize the steps to be taken to develop a reversed-phase separation for more complex mixtures (e.g., 10–20 components) when gradient elution is available; method development then becomes more convenient and proceeds faster. The final units in this chapter (Sects. 9.4, 9.5) detail the steps to be taken to develop isocratic *normal-phase* and *ion-pair* separations.

All of these procedures are designed so that initial step(s) provide the greatest potential for developing a successful final method. Adequate condi-

227

TABLE 9.1   Flowsheet of Reversed-Phase HPLC Method Development

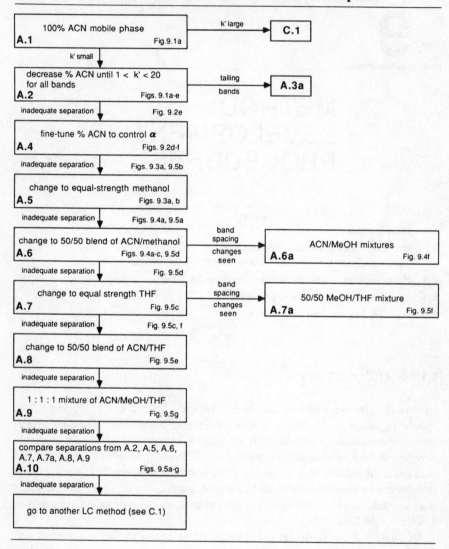

TABLE 9.2 Flow-Sheet of Normal-Phase HPLC Method Development (From Sect. 9.4)

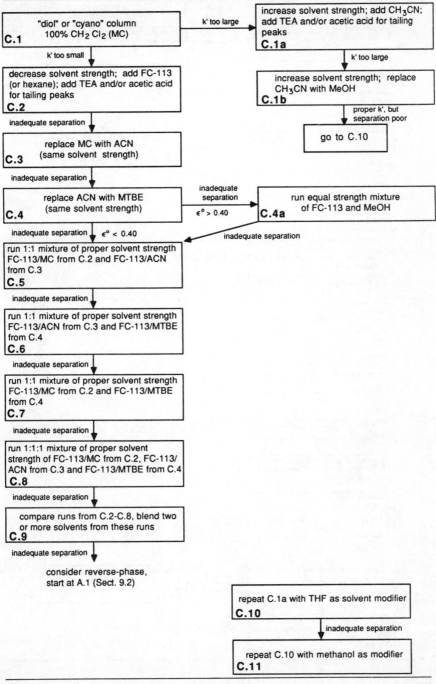

TABLE 9.3 Flow-Sheet of Ion-Pair HPLC Method Development (From Sect. 9.5)

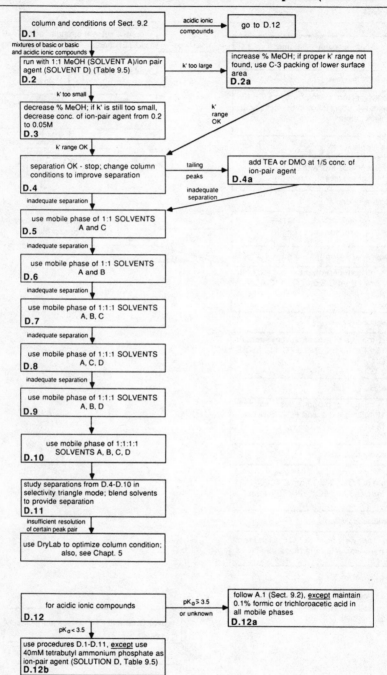

tions for easy separations often can be found with only a few experimental runs. As the separation of interest becomes progressively more difficult, it is likely that more experimental runs will be required. *It should be emphasized that if an adequate separation for the problem to be solved is found at any step in the procedure (i.e., required resolution, satisfactory run time, acceptable column back pressure), method development is complete; no further work is required.*

It is especially important to note that at any step during method development, column conditions (flow rate, column length, particle size) can be optimized to produce a better separation, according to the concepts in Sect. 2.5, or by using computer simulation (DRYLAB, Sect. 8.4).

Certain preliminary conditions must be established before starting method development. Table 9.4 lists various factors (together with pertinent reference sections) that should be considered before undertaking the development of a particular separation method.

The approach given in the following sections uses changes in solvent strength and solvent selectivity to provide the variations in band spacings necessary to produce the required separation. With each change in mobile phase, it is assumed that the column will be appropriately equilibrated (e.g., about 20 column volumes) with the new solvent, before the next separation is attempted.

Illustrative chromatograms are used to demonstrate each of the more important steps in the following reversed-phase method-development approaches. These examples can be used as guides for developing the separation of other samples.

9.2 ISOCRATIC REVERSED-PHASE SEPARATIONS

General Conditions for All Reversed-Phase Separations

Use a 15- or 25-cm $\times$ 0.46-cm column (C-8 or C-18, 8–10 nm pores; see Tables 3.1 and 3.2); set column temperature at 40° C (if possible); for acidic or basic samples, add 0.05 M phosphoric acid to the water in the mobile phase (adjust to pH 3.5 with sodium hydroxide).

A.1

Equilibrate the column with 100% acetonitrile at a flow rate of 1.5 mL/min, until a stable detector baseline is obtained. Inject 10 μL of a 1 mg/mL, 1/1 acetonitrile/water solution of the sample mixture into the column (sample

TABLE 9.4   Factors to Consider before Developing the Separation

Item	Reference Sections
• Nature of sample determined (suggests a particular method)	1.1, 1.2
• Separation goal defined	1.1
• Required resolution established	2.1
• Detector selected	1.1, 4.1
• Sample size established	4.1

size may have to be varied according to the discussion in Sect. 4.1). Should all components elute at or near t_0 (about 1 min; see Fig. 9.1a), go to A.2. If no peaks are seen after about 10 min, go to C.1 (Sect. 9.4).

A.2

Decrease the percent acetonitrile in the mobile phase stepwise by diluting with water in 20%v increments (i.e, acetonitrile = 80%, 60%, as in Figs. 9.1b,c) until the last peak of the mixture elutes at about $k' \geq 1$ ($t_R = 2$ min; Fig. 9.1c). Continue to decrease the percent acetonitrile in 10%v steps (threefold k' rule of Sect. 2.2)—until peaks of the mixture elute within a range of $1 < k' < 20$ (Sect. 4.2) (Figs. 9.1d,e). If the k'-values for peaks can be adjusted into this range by changing the percent acetonitrile, proceed to A.3. Should the sample have a k'-range of more than 20, gradient elution will be required; go to B.1 (Sect. 9.3). If the peaks cannot be adjusted into the desired range of $1 < k' < 20$ (e.g., with 5% acetonitrile), another HPLC method may be required; go to C.1 (Sect. 9.4).

A.3

If a satisfactory separation is obtained ($R_s \geq 2$ for all band pairs as in Fig. 9.1e), the method is finished. Should tailing peaks be found at any step, go to A.3a. If one or more peak pairs of interest are not adequately separated after adjusting the mobile-phase strength to produce $k' = 1 - 20$ (as in the sequence for Figs. 9.2a–e), proceed to A.4.

A.3a.   For a mixture of acidic and basic compounds, add 30 mM triethylamine acetate (TEA/AC) to the mobile phase to eliminate band tailing. Should an adequate separation result, stop. If tailing is evident for bands of basic compounds, make sure that a basic column is being used (see Sect. 3.1). If tailing persists, add 10 mM of dimethyloctylamine to the mobile phase (in-

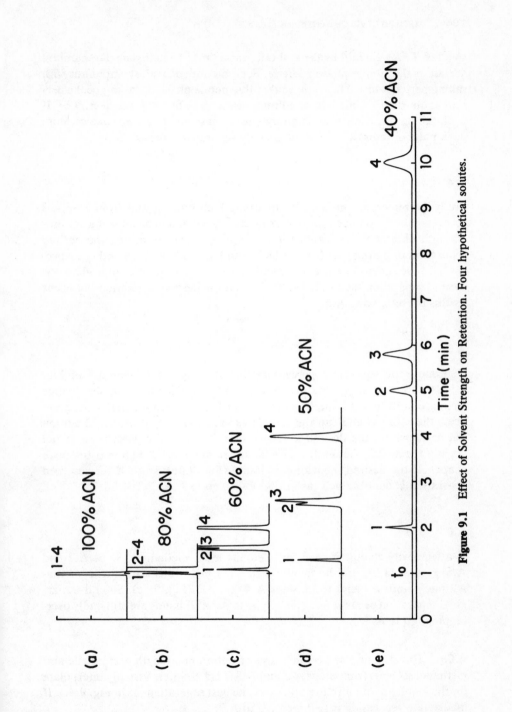

Figure 9.1 Effect of Solvent Strength on Retention. Four hypothetical solutes.

233

stead of TEA). Should peaks still tail, use 20 mM trimethyloctylammonium acetate as the mobile-phase additive. For a mixture of acidic compounds with tailing peaks, add 1.0% acetic acid to the mobile phase. If tailing peaks persist, another HPLC method or column will have to be used (see Sect. 9.5). If peak tailing has been eliminated by one of these additives, but one or more peak pairs of interest are not adequately separated, proceed to A.4.

A.4

Study the retention time vs. solvent strength chromatograms from Step A.3 (process may have to be repeated as in Step A.3a) and determine if a solvent-strength difference can cause the overlapping peak(s) limiting the desired separation to move apart (see peaks 2,3 and 4,5 in Figs. 9.2d and e, respectively). If so, determine the intermediate percent acetonitrile to produce the desired separation (as in Fig. 9.2f). Should the desired separation not occur by this process, go to A.5.

A.5

Determine the equivalent percent methanol using the nomograph of Fig. 2.14, to produce an approximately equal k' range for the sample. Change from acetonitrile (inadequate separation of Fig. 9.3a) to methanol, reequilibrate the column, and run the sample as in A.1 (Fig. 9.3b). Should elution not occur within the desired k' range of 1–20, alter the percent methanol slightly (Sect. 2.2) (maintain TEA or other modifier, if required for peak shape). If the desired separation is found (Fig. 9.3b), stop. If the required separation is not obtained (as in Fig. 9.4b), go to A.6.

A.6

If a selectivity change is seen for acetonitrile vs. methanol, (Steps A.2 and A.5; Figs. 9.4a,b), run the sample with a 1/1 mixture of acetonitrile/methanol mobile-phase blend from Steps A.4 and A.5 (Fig. 9.4c). Should a marginal separation occur (Figs. 9.4a–c), go to A.6a. If bands are still badly overlapping, go to A.7.

A.6a. Run 1/3 and 3/1 (v) mixtures of optimum-strength acetonitrile and methanol solvents from Steps A.2 and A.5 (Figs. 9.4d,e). Visually interpolate for the mobile-phase mixture that gives the best separation (as in Fig. 9.4f). If inadequate separation is still seen, go to A.7.

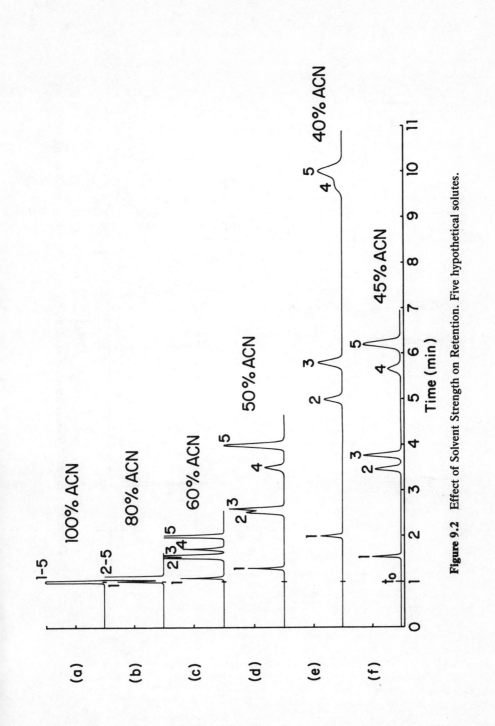

Figure 9.2 Effect of Solvent Strength on Retention. Five hypothetical solutes.

235

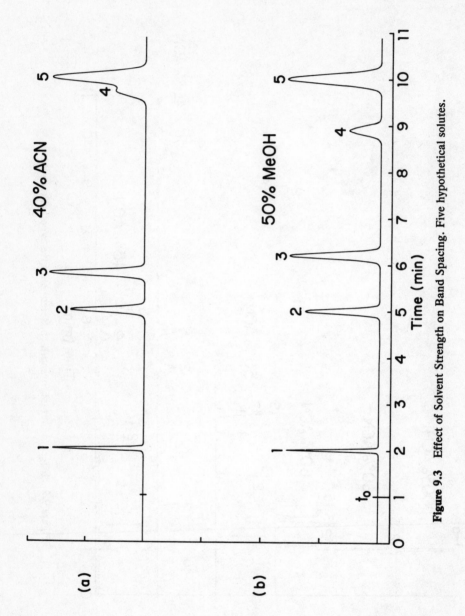

Figure 9.3 Effect of Solvent Strength on Band Spacing. Five hypothetical solutes.

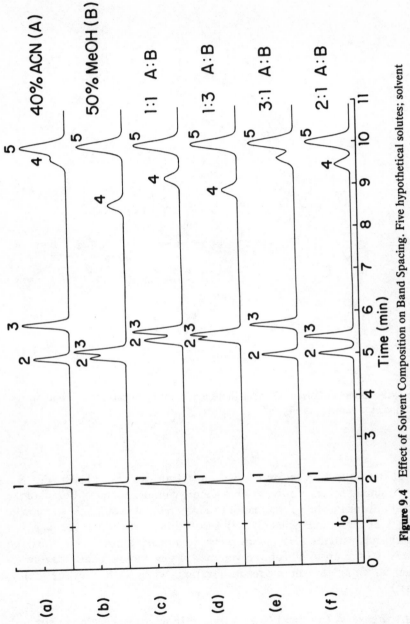

Figure 9.4 Effect of Solvent Composition on Band Spacing. Five hypothetical solutes; solvent A, 40% acetonitrile; solvent B, 50% methanol.

237

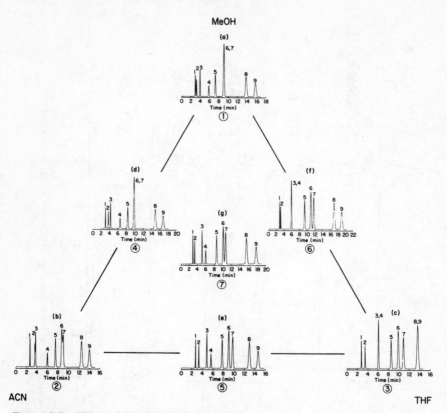

Figure 9.5 Effect of Solvent Composition on Separating Substituted Naphthalenes. Conditions the same as in Fig. 8.9.

A.7

Repeat the separation of Step A.5 using tetrahydrofuran (THF) (Fig. 9.5 shows a similar scheme for separating a more complex sample). Compare the separations for Position 1/methanol in Fig. 9.5a and Position 2/acetonitrile in Fig. 9.5b with the Position 3/THF separation in Fig. 9.5c). If changes in band spacing with the THF mobile phase are inadequate (as in Fig. 9.5c), go to A.7a. (Note: up to 40-column volumes of new mobile phase may be required for complete column reequilibration, when THF is in the mobile phase.)

A.7a. Repeat A.6a using 1/1 methanol/THF mixture of optimum-strength solvents from Steps A.5 and A.7 (Position 6, Fig 9.5f). If the desired separation is not obtained (as in Fig. 9.5f), go to A.8.

A.8

Run the sample with a 1/1 mixture of optimum-strength acetonitrile/THF solvents from Steps A.2 and A.7 (see Position 5, Fig. 9.5e). If the desired separation for the mixture is not obtained, go to A.9. (Should a particular pair of known components not be separated by the steps to this point, the proposed reverse-phase method is not likely to succeed for this pair; do not proceed to A.9. Another HPLC method should be attempted [Sects. 9.4 or 9.5]).

A.9

If the desired separation has not been obtained, run the sample with a 1/1/1 mixture of equal-strength acetonitrile/methanol/THF solvents from Steps A.2, A.5, and A.7 (maintain the TEA or other modifiers as required) (Position 7, Fig. 9.5g). If the separation still is not adequate, go to A.10.

A.10

Visually compare the seven chromatograms from Steps A.2, A.5, A.6, A.7, A.7a, A.8, and A.9 as in Fig. 4.5 for Positions 1–7, respectively (see chromatograms for the eight-component mixture in Fig. 9.5). Appropriately blend solvents from these positions (see Sect. 4.3) to produce the desired separation. (Note: optimum separation of the eight-component mixture actually is obtained with a ternary solvent mixture close to Position 5, Fig. 9.5e). If desired separation is not obtained in this way, another HPLC method may be required. Go to C.1 (Sect. 9.4).

9.3   REVERSED-PHASE SEPARATIONS BY GRADIENT ELUTION

B.1

Use the column and initial conditions of Sect. 9.2. Purge the column with at least 20 column volumes of 100% acetonitrile at a flow rate of 1.0 mL/min, until a stable detector baseline is obtained. Equilibrate the column with 20 column volumes of 5% acetonitrile/water. Inject 10 μL of a 1 mg/mL, 1/1 acetonitrile/water solution of the sample mixture (sample size may have to be suitably adjusted here or later during method development, according to the discussion in Sect. 4.1). Run a 20-min linear gradient of 5–100% acetonitrile. If all peaks elute at or near t_0, another HPLC method will be required. (If ionic compounds are involved, consider ion pairing as in Sect. 9.5. For non-ionic compounds, select normal-phase chromatography with polar mobile phases as in Sect. 9.4). Should the sample elute after the gradient is completed, use a "weaker" column (i.e., C-3, C-4, cyano, and/or lower surface areas, e.g., particles with ≥ 30-nm pores). If tailing peaks are seen, add modifiers to the aqueous mobile-phase component, as described in Step A.3a, Sect. 9.2.

Use the relationship shown in Fig. 9.6a to confirm that gradient elution should be used for the sample. If the difference between the retention times of the first (t_i) and last (t_f) peaks of the chromatogram (Δt_g) divided by the gradient time t_G is >0.25, a gradient is needed. If the peaks of the mixture eluting during the gradient are not adequately separated (as in Fig. 9.6a), go to B.9. If $\Delta t_g < 0.25$, an isocratic separation is indicated. Go to B.3.

B.2

Use the 20-min gradient run in B.1 (example in Fig. 9.6a) and the nomograph in Fig. 9.7 to estimate the percent acetonitrile required to isocratically elute the sample in the desired range $1 < k' < 20$. (For example, if $t_x = 16.0$, for a 15-cm column and a one-pump gradient system, Fig. 9.7 predicts an isocratic mobile phase containing 41% ACN). Adjust the percent acetonitrile, if required. If the resulting separation is inadequate, run chromatograms with 5%v more and 5% less acetonitrile. Should a solvent-strength difference cause useful changes in band spacings (as in Figs. 9.2d,e), determine the percent acetonitrile that produces the desired separation. If the separation is inadequate, go to B.3.

B.3–B.8

These steps are the same as A.5–A.10 in Sect. 9.2.

B.9

For samples with a k'-range of >20, gradient elution will be required for all runs. The first step is to optimize the gradient range by adjusting the initial percent acetonitrile (5% acetonitrile minimum) for (a) minimum separation time for the initial peak at t_i (Fig. 9.6a), and (b) adequate spacing of the early-eluting peaks from each other and from t_0 (see Sect. 6.3). Next, the final percent acetonitrile is selected to elute the last peak of the mixture at t_f (Fig. 9.6a) near the end of the gradient, as discussed in Sect. 6.3. With the 20-min gradient run of B.1, use the nomographs in Fig. 9.8 to estimate the initial and final values of percent acetonitrile for an optimized run. (For example, with a

15-cm column, a one-pump gradient system and for $t_i = 9.5$ and $t_f = 18.0$ [as in Fig. 9.6a], Fig. 9.8 predicts an initial mobile phase with 19% ACN and a final concentration of 72% for this 20-min gradient.) Adjust mobile phase strength, if required. Should an inadequate separation of desired peaks occur (as in Fig. 9.6b), go to B.10.

B.10

Run the sample with a 10-min linear gradient using the initial and final percent acetonitrile values from B.9. If an inadequate separation is seen, go to B.11.

B.11

Run the sample with a 40-min linear gradient using the initial and final percent acetonitrile values from B.9 (as in Fig. 9.6c). If one or more peak pairs remains unresolved, go to B.12.

B.12

If useful peak-spacing changes appear between the 20-min gradient of B.1 and the 10-min gradient of B.10, run a 15-min gradient. If one or more peak pairs remain unresolved, and there are useful peak-spacing changes between the 20-min gradient of B.1 and the 40-min gradient of B.11, run a 30-min gradient with the same initial and final percent acetonitrile. Interpolate for an optimum gradient time for the desired separation. If inadequate separation is still seen, go to B.13.

B.13

Repeat Steps B.9–B.12 using methanol as the solvent. Use the nomograph in Fig. 2.14 to estimate the initial and final percent methanol required for these gradient runs, based on the percent acetonitrile previously used. If inadequate separation occurs, go to B.14.

B.14

Repeat Step B.13 using tetrahydrofuran as the solvent. If inadequate separation is seen, another method may be required. Go to Sect. 9.4.

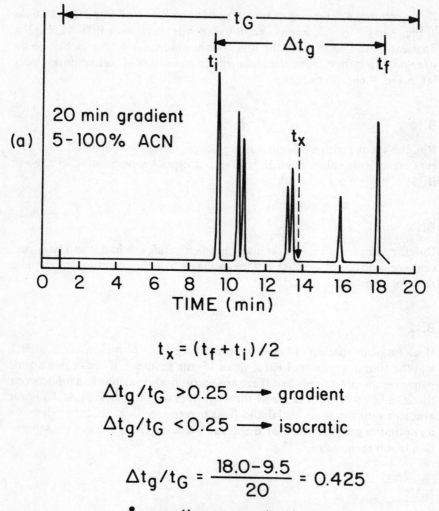

$$t_x = (t_f + t_i)/2$$

$$\Delta t_g/t_G > 0.25 \longrightarrow \text{gradient}$$

$$\Delta t_g/t_G < 0.25 \longrightarrow \text{isocratic}$$

$$\Delta t_g/t_G = \frac{18.0 - 9.5}{20} = 0.425$$

∴ gradient required

Figure 9.6 Determining Whether to Use Isocratic or Gradient Runs; Effect of Gradient Time. (a) Determining isocratic or gradient; (b,c) effect of gradient time on separating seven hypothetical solutes. See text in Sect. 9.3.

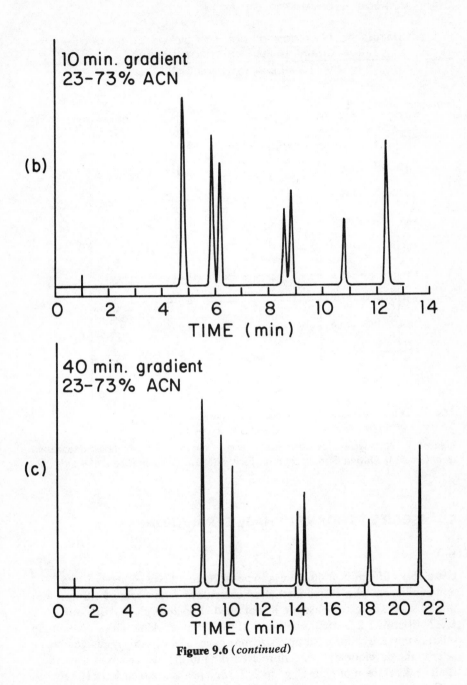

Figure 9.6 (*continued*)

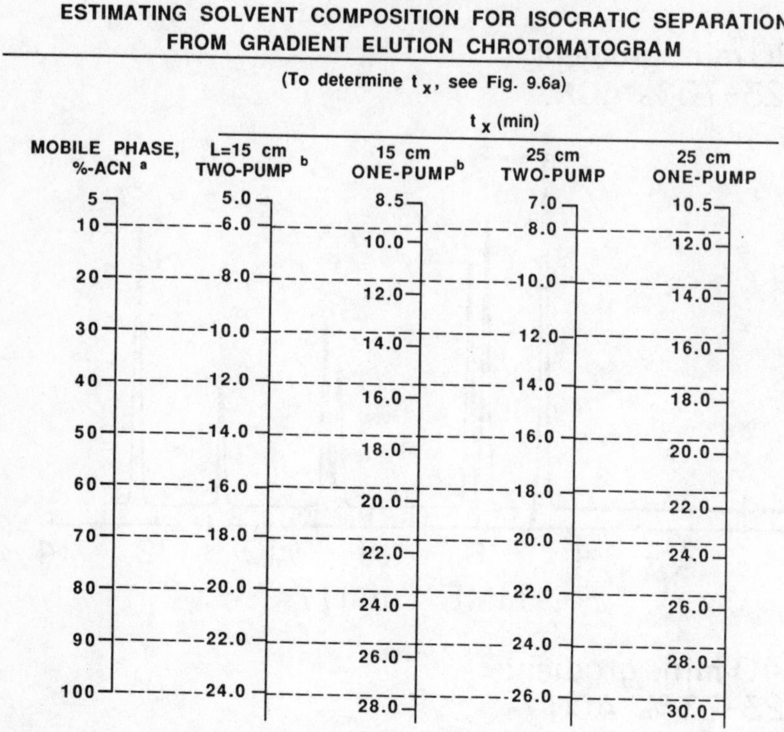

ESTIMATING SOLVENT COMPOSITION FOR ISOCRATIC SEPARATION FROM GRADIENT ELUTION CHROTOMATOGRAM

(To determine t_x, see Fig. 9.6a)

t_x (min)

MOBILE PHASE, %-ACN [a]	L=15 cm TWO-PUMP [b]	15 cm ONE-PUMP[b]	25 cm TWO-PUMP	25 cm ONE-PUMP
5	5.0	8.5	7.0	10.5
10	6.0	10.0	8.0	12.0
20	8.0	12.0	10.0	14.0
30	10.0	14.0	12.0	16.0
40	12.0	16.0	14.0	18.0
50	14.0	18.0	16.0	20.0
60	16.0	20.0	18.0	22.0
70	18.0	22.0	20.0	24.0
80	20.0	24.0	22.0	26.0
90	22.0	26.0	24.0	28.0
100	24.0	28.0	26.0	30.0

[a] See Fig. 2.14 to convert %-ACN to other mobile phase solvents.

[b] Gradient system; dwell volume, 2.0 mL and 5.5 mL for two- and one-pump mixing, resp.

Figure 9.7 Nomograph for Estimating Solvent Composition for Isocratic Separation from Gradient Elution Chromatogram. Flow rate, 1.0 mL/min. See Fig. 9.6a.

9.4 ISOCRATIC NORMAL-PHASE SEPARATIONS

C.1

Use a 15- or 25-cm $\times$ 0.46-cm cyano or diol column (see Tables 3.1 and 3.2), and equilibrate with 100% methylene chloride at a flow rate of 1.5 mL/min until a stable detector baseline in obtained. Inject 10 μL of 1 mg/mL, 1/1 1,1,2-trifluoro-1,2,2-trichloroethane (FC-113)/methylene chloride sample mixture into the column (sample size may have to be suitably adjusted here or later in the development procedure according to the discussion in Sect. 4.1). If all peaks elute at or near t_0, go to C.2. If no peaks are seen after 10 min, go to C.1a.

ESTIMATING OPTIMUM INITIAL AND FINAL SOLVENT COMPOSITIONS
FOR 20 MIN GRADIENT

(Conditions of step B.1; see Fig. 9.6a for measuring t_i and t_f)

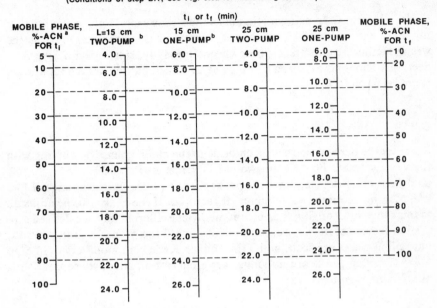

[a] See Fig. 2.14 to convert %-ACN to other mobile phase solvents.

[b] Gradient system; dwell volume, 2.0 mL and 5.5 mL for two- and one-pump mixing, resp.

Figure 9.8 Nomograph for Estimating Optimum Initial and Final Solvent Compositions for 20-min Gradient. Flow rate, 1.0 mL/min. See Fig. 9.6a and text in Sect. 9.3.

C.1a.   Increase the strength of the mobile phase by adding acetonitrile; initially try a 1/1 mixture of methylene chloride/acetonitrile. If the peaks elute at or near t_0, decrease the acetonitrile content by successively halving the concentration, until the peaks of the mixture elute in a range of $1 < k' < 20$; make minor solvent-strength adjustments, if required. Should no peaks appear after 10 min with the 1/1 methylene chloride/acetonitrile mixture, increase the concentration of acetonitrile until the proper k' range of 1–20 is obtained. If this k' range is not obtained with any concentration of acetonitrile, go to C.1b. If the desired k' range results, but tailing peaks are apparent, add 0.1% TEA (or dimethyloctylamine) to eliminate tailing of basic compounds, 0.1% acetic acid to eliminate tailing of acidic compounds, or 0.1% of each if a mixture of both basic and acidic compounds is involved (a slight adjustment in acetonitrile concentration may have to made, to maintain proper k′ range). If the proper k′ range is obtained with a methylene chloride/acetonitrile mixture, but the desired separation is not seen, go to C.10.

C.1b.   Repeat C.1a using methanol as the strong mobile-phase modifier. If no peaks elute with 100% methanol, normal-phase chromatography is the wrong method; try reversed-phase chromatography, starting with Step A.1, Sect. 9.2. Should the desired k′ range be obtained, but the desired separation is not seen, go to C.10.

C.2

Decrease the percent methylene chloride in the mobile phase by diluting with the weak solvent, FC-113, in twofold increments (i.e., 50%, 25%, 12.5% methylene chloride, etc.) until the peaks of the mixture elute in a range of $1 < k′ < 20$; make minor adjustments, if required. If the peaks can be adjusted to this range by changing the percent methylene chloride, but the desired separation is not obtained, proceed to C.3. Should the proper k′ range result, but tailing peaks are seen, add TEA and/or acetic acid modifiers as in Step C.1a. If tailing peaks are no longer apparent but an adequate separation is not obtained, go to C.3.

C.3

Run an equal-strength mixture of the weak solvent FC-113 and acetonitrile (ACN) (instead of methylene chloride; maintain proper k′ range and the TEA/acetic acid additives as in C.1a, if required); see Table 2.2 for solvent-strength values. If the desired separation is not obtained, go to C.4.

C.4

Run an equal-strength mixture of FC-113 and methyl-t-butyl ether (MTBE) instead of acetonitrile, as in Step C.3; see Table 2.2 for solvent-strength values. If the desired separation is not obtained, and the solvent-strength value required for the proper k′ range is $\epsilon° > 0.40$, go to C.4a. If the desired separation is not obtained and the solvent strength value required for the proper k′ range is $\epsilon° < 0.40$, go to C.5.

C.4a.   Using Table 2.2 as a guide, run an equal-strength mixture of FC-113 and methanol (instead of FC-113 and ACN), as in C.3 (if hexane is used as the weak solvent instead of FC-133, isopropanol must be used as the modifier to insure solvent miscibility). If the desired separation is not found, go to C.5.

C.5

Run a 1/1 mixture of the proper-strength mobile phase (FC-113/MC) from C.3 and the corresponding mobile phase (FC-113/ACN) from Step C.4. If the desired separation is not obtained, go to C.6.

C.6

Run a 1/1 mixture of the proper-strength mobile phase (FC-113/ACN) from Step C.3 and the corresponding mobile phase (FC-113/MTBE) from Step C.4. If the desired separation is not found, go to C.7.

C.7

Run a 1/1 mixture of the proper-strength mobile phase (FC-113/MC) from Step C.2 and the corresponding mobile phase (FC-113/MTBE) from Step C.4. If the desired separation is not obtained, go to C.8.

C.8

Run a 1/1/1 mixture of the proper-strength mobile phase (FC-113/MC, FC-113/ACN, and FC-113/MTBE) from Steps C.2, C.3, and C.4, respectively. If the desired separation is not found, go to C.9.

C.9

Compare the seven chromatograms from the C.2–C.8 runs. Appropriately blend two or more of these mobile phases (see Sect. 4.3) to produce the desired separation. If the desired separation is not obtained, consider a reversed-phase separation starting at A.1 (in cases where some, but inadequate, normal-phase separation is seen). For especially difficult separations, see Chapter 5.

C.10

Repeat step C.1a using tetrahydrofuran (THF) as the mobile-phase modifier (slightly adjust the MC/THF ratio, if needed to maintain proper k' range). If the desired separation is not found, go to C.11.

C.11

Repeat Step C.10 using methanol (or isopropanol, if hexane is the base solvent).

9.5 ION-PAIR SEPARATIONS

D.1

Use the column and conditions of Sect. 9.2. To separate basic ionic compounds, go to D.2; to separate acidic ionic compounds, go to D.12; to separate mixtures of basic and acidic compounds, go to D.2.

D.2

Equilibrate the column with a mobile phase of one part methanol and one part ion-pairing agent (SOLVENT D) (see Table 9.5 for a listing of SOLVENTS) at a flow rate of 1.5 mL/min, until a stable detector baseline is obtained. Inject 10 μL of a 1 mg/mL mobile-phase solution of the sample mixture into the column (sample size may have to be suitably altered here or later in the procedure; see Sect. 4.1). If all components elute at or near t_0, go to D.3. If no peaks are seen after about 10 min, go to D.2a.

D.2a. Increase the percent methanol in the mobile phase in stepwise fashion by 10% increments (i.e., 60%, 70%, etc.; maintain the other mobile-phase components), until the peaks of the mixture elute in the range of $1 < k' < 20$ (Sect. 4.2). If the peaks are adjusted to this range by a particular methanol concentration (SOLVENT A), go to Step D.4; if not, repeat Step

TABLE 9.5 **Suggested Mobile Phases for Ion-Pairing HPLC**

Mixture	Type	Composition[a]
SOLVENT A	Methanol	Methanol
SOLVENT B	Buffer pH 2.5	50 mL 0.05 M phosphoric acid + 50 mL 0.05 M acetic acid; adjust to pH 2.5 with 2 M sodium hydroxide ($\sim$1.5 mL)
SOLVENT C	Buffer pH 7.0	50 mL 0.05 M phosphoric acid + 50 mL 0.05 M acetic acid; adjust to pH 7.0 with 2 M sodium hydroxide ($\sim$5.7 mL)
SOLVENT D	Ion-pairing agent/buffer[b]	0.20 M hexane sulfonate in buffer (50 mL 0.005 M phosphoric acid + 50 mL 0.005 M acetic acid; adjust to pH 5.0 with 0.2 M sodium hydroxide [$\sim$3.5 mL])

[a]May be necessary to add 20 mM TEA to all SOLVENTS.
[b]For separating acids, use 0.04 M tetrabutylammonium phosphate (instead of hexane sulfonate) as ion-pairing agent. With a detector wavelength of <220 nm, acetic acid must not be present; phosphate buffers are tolerated.

D.1 using a C-3 or C-4 reversed-phase packing with 30–50 nm pores (lower surface area).

D.3

Decrease the percent methanol in the mobile phase in stepwise fashion by 10%v increments (i.e., 40%, 30%, etc.; maintain the other mobile phase components), until the peaks of the mixture elute in a range of $1 < k' < 20$ (see Sect. 4.2). If the peaks are adjusted to this range with this methanol concentration (SOLVENT A), go to D.4 (if not, repeat Step D.2 using 0.05 M hexane sulfonate ion-pairing agent, instead of 0.2 M [Table 9.5], to further reduce peak retentions).

D.4

If the desired separation is obtained for the methanol concentration providing the proper k' range (SOLVENT A), stop. Should tailing peaks be found for a mixture of basic compounds, go to D.4a. If one or more peak pairs of interest are not adequately separated, go to D.5.

D.4a.   For the tailing of basic samples, add triethylamine (TEA) (or dimethyloctylamine) in an amount up to one-fifth the concentration of ion-pair reagent. If the resultant separation is not adequate, go to D.5.

D.5

Use a mobile phase consisting of one part SOLVENT A and one part SOLVENT C (see Table 9.5). (Note: when concentration of mobile phase components is changed, a minor change in percent methanol may be required to maintain proper k' range.) Should the desired separation not occur, go to D.6.

D.6

Use a mobile phase of one part SOLVENT A and one part SOLVENT B (see Table 9.5, SOLVENT B). If the separation is inadequate, go to D.7.

D.7

Use a mobile phase of one part each of SOLVENTS A, B, and C (see Table 9.5). If the separation is inadequate, go to D.8.

D.8

Use a mobile phase of one part each of SOLVENTS A, C, and D (see Table 9.3). If the separation is inadequate, go to D.9.

D.9

Use a mobile phase of one part each of SOLVENTS A, B, and D (see Table 9.5). If the separation is inadequate, go to D.10.

D.10

Use a mobile phase of one part each of SOLVENTS A, B, C, and D (see Table 9.5). If the separation is inadequate, go to D.11.

D.11

Place the chromatograms from Steps D.5, D.4, and D.3 at the top, left, and right corners of a triangle, respectively, for Positions 1, 2, and 3 corresponding to three binary mixtures with methanol (see Fig. 4.12). The chromatogram from Step D.7 is placed midway at Position 4 between Positions 1 and 2; the chromatogram from Step D.8 is placed at Position 5, midway between Positions 2 and 3; the chromatogram from Step D.9 is placed at Position 6, midway between Positions 1 and 3; and, the chromatogram from Step D.10 is placed in the center of the triangle at Position 7. (See Figure 4.13 for an example of the configuration of these chromatograms.) Study the changes in band spacings for the various positions and select a solvent blend (by visual interpolation) that will provide the separation of interest. If insufficient resolution is seen for the peak pair(s) of interest, consider the use of the computer-assisted DRYLAB approach of Sect. 8.4. For especially difficult separations, see Chapter 5.

D.12

For mixtures of acidic ionic compounds, two different approaches are feasible. For mixtures of compounds with pK_a-values of about 3.5 or greater, go to D.12a; for more acidic mixtures, go to D.12b. If the level of acidity is unknown, first try D.12a, then, if unsuccessful, try D.12b.

D.12a.   Go to A.1 in Section 9.2 and follow procedure, *except* maintain 0.1% trifluoroacetic acid in all mobile-phase mixtures.

D.12b.   Use same procedures as given in Steps D.1–D.10, *except* use 40 mM tetrabutylammonium phosphate as the ion-pairing agent (as SOLVENT D for acids in this case).

INDEX